Museums and the History of Computing

IØØ31226

Museums and the History of Computing examines the critical role that cultural organizations, such as museums and galleries, play in shaping 'digital heritage': the cultural heritage surrounding computer technology.

Focusing on digital technologies as objects and practices that museums collect, exhibit, and preserve for the future, this book highlights how and why museums play a crucial role in preserving the rich heritage of the digital world, constructing powerful narratives that help make it relevant to the public. It demonstrates that the museum can be a powerful means of safeguarding and interpreting ephemeral and continually changing digital technology, offering new pathways for rethinking the very meaning of digital objects and practices in contemporary societies. It provides practices and strategies for the preservation and exhibition of computing artifacts and ways to accommodate and respond to narratives about histories of computing that circulate in the public arena. Bringing together leading museum and university researchers and practitioners, and mobilizing cross-cutting debates and approaches in areas such as museum studies, cultural heritage, history of technology, anthropology, and media studies, this book challenges us to think critically about what 'digital' is when examined not only as a tool but as a cultural object deserving of attention and a place within the museum.

Museums and the History of Computing is for museum studies students and researchers as well as museum practitioners – especially those with an interest in digital technology and heritage. It will be of interest to researchers and students interested in histories of computing and digital media and in digital media studies.

Simone Natale is an associate professor at the University of Turin, Italy, and an editor of *Media, Culture and Society*. He is the author of *Deceitful Media: Artificial Intelligence and Social Life after the Turing Test* (Oxford University Press, 2021).

Petrina Foti is a museologist and scholar focused on the rise of digital information and technology and the resulting impact on both museums and the wider world. She is the author of *Collecting and Exhibiting Computer-based Technology: Curatorial Expertise at the Smithsonian Museums* (Routledge, 2018).

Ross Parry is a professor of museum technology at the University of Leicester, and the inaugural Director of its Institute for Digital Culture. He is co-founder the UK's Museum Data Service, and co-editor of *The Routledge Handbook of Museums, Media and Communication* (Routledge 2019).

Critical Perspectives on Museums and Digital Technology
Series Editors: Vince Dziekan and Ross D. Parry

Critical Perspectives on Museums and Digital Technology contends that digital – now more than ever – is critical to the future of the cultural sector, but that it is also a relationship in need of urgent reframing.

Promoting interdisciplinary writing that seeks to address contemporaneous issues in a time-sensitive way, the series will also be perceptive and attuned to contextual precedents and emergent issues from across the museum technology field. Contributions will be "critical" in that each book will always bring a high level of criticality and intellectuality to the way it negotiates its subject. These publications will also speak to issues that are timely, urgent and, therefore, "critical" to how we understand current practices. As a whole, the series will consider how to redefine the museum in relation to the sociocultural and technological conditions within which it must coexist.

Critical Perspectives on Museums and Digital Technology will include solo, collaborative and ensemble volumes that are accessible and supported by practice-based insights. The series will be essential reading for academics, students and professionals around the globe. Contributions to the series will be particularly useful to those with an interest in museums and heritage, media and communication, art and design, education, business and economics.

Museums and the History of Computing
Objects, Narratives and Practice
Simone Natale, Petrina Foti and Ross Parry

The following list includes only the most-recent titles to publish within the series. A list of the full catalogue of titles is available at: www.routledge.com/Critical-Perspectives-on-Museums-and-Digital-Technology/book-series/CPMDT

Museums and the History of Computing

Objects, Narratives and Practice

**Edited by Simone Natale,
Petrina Foti and Ross Parry**

Routledge
Taylor & Francis Group

LONDON AND NEW YORK

First published 2025
by Routledge
4 Park Square, Milton Park, Abingdon, Oxon OX14 4RN

and by Routledge
605 Third Avenue, New York, NY 10158

Routledge is an imprint of the Taylor & Francis Group, an informa business

© 2025 selection and editorial matter, Simone Natale, Petrina Foti,
Ross Parry; individual chapters, the contributors

The right of Simone Natale, Petrina Foti, Ross Parry to be identified as
the authors of the editorial material, and of the authors for their individual
chapters, has been asserted in accordance with sections 77 and 78 of the
Copyright, Designs and Patents Act 1988.

All rights reserved. No part of this book may be reprinted or reproduced or
utilised in any form or by any electronic, mechanical, or other means, now
known or hereafter invented, including photocopying and recording, or in any
information storage or retrieval system, without permission in writing from the
publishers.

Trademark notice: Product or corporate names may be trademarks or
registered trademarks, and are used only for identification and explanation
without intent to infringe.

British Library Cataloguing-in-Publication Data
A catalogue record for this book is available from the British Library

ISBN: 978-1-032-54401-4 (hbk)
ISBN: 978-1-032-54402-1 (pbk)
ISBN: 978-1-003-42470-3 (ebk)

DOI: 10.4324/9781003424703

Typeset in Times New Roman
by Apex CoVantage, LLC

Contents

Figures

Contributors

Petrina Foti is a museologist and scholar focused on the rise of digital information and technology and the resulting impact on both museums and the wider world. She is the author of *Collecting and Exhibiting Computer-based Technology: Curatorial Expertise at the Smithsonian Museums* (Routledge, 2018) and holds the honorary position of Science Museum Group Research Associate (SMGRA) at the Science Museum in London (UK).

Simone Natale is an associate professor at the University of Turin, Italy, and an editor of *Media, Culture and Society*. Before taking up in 2020 a position in Turin, his hometown, he taught and researched at Columbia University in New York City, US; Concordia University in Montreal, Canada; Humboldt University Berlin and the University of Cologne in Germany; and Loughborough University in the UK. He is the author of two monographs, including *Deceitful Media: Artificial Intelligence and Social Life after the Turing Test* (Oxford University Press, 2021), which has been translated into Italian, Chinese, and Portuguese, as well as more than forty articles published in leading peer-reviewed journals such as *New Media & Society, Communication Theory*, the *Journal of Communication, Convergence*, and *Information, Communication & Society*. His research has been funded by international organizations, including the AHRC and the ESRC in the UK, MIUR in Italy, the Humboldt Foundation and the DAAD in Germany, and Columbia University's Italian Academy in the US.

Ross Parry is a principal fellow of the Higher Education Academy, a former Tate Research fellow, and former chair of the UK's national Museums Computer Group. He is one of the founding Trustees of the Jodi Mattes Trust – for accessible digital culture. In 2018 he was listed in the Education Foundation's 'EdTech50' – the 50 most influential people in the UK education and technology sectors. Ross served on the International Scientific Advisory Board for 'Learning 2.0' managed by DREAM (the Danish Research Centre on Education and Advanced Media Materials) at the University of Southern Denmark, where in 2012 he was visiting professor. He now sits on the International Advisory Board for the €6mn, five-year, 'Our Museum' project, funded by Nordea-Fonden and Velux Fonden, as well as the UK Research and Industry's Steering Committee of its £19mn digital cultural heritage initiative 'Towards a National Collection.' His recent books include: *Museum Thresholds: The Design and Media of Arrival*, edited with Ruth Page and Alex Moseley (Routledge, 2018), and *The Routledge Handbook of Media and Museums* (2019), edited with Kirsten Drotner, Vince Dziekan, and Kim Schrøder. Ross is the author of *Recoding the Museum: Digital Heritage and the Technologies of Change* (Routledge 2007), and in 2010 published *Museums in a Digital Age* (also with Routledge). Ross leads the 'One by One' international consortium of museums, professional bodies, government agencies, commercial partners, and academics that together are working to build digitally confident museums. After a three-year national project in the UK (working with the Museums Association, Arts Council, and National Lottery Heritage Fund), the consortium's latest project ('structuring museums to deliver new digital experiences') now brings partners including the V&A, Science Museum, and the UK's Museums Computer Group into an action research collaboration with the Smithsonian Institution, American Alliance of Museums, and Museum Computer Network.

Joshua Bell is cultural anthropologist and the Curator of Globalization at the Smithsonian Institution's National Museum of Natural History. He also co-directs Recovering Voices' Mother Tongue Film Festival and runs the Summer Institute in Museum Anthropology (SIMA). Combining ethnographic fieldwork with research in museums, Bell examines the shifting local and global network of relationships between persons, artefacts and the environment. To date, this has involved him carrying out research in Papua New Guinea, Hawai'i, and the Washington, DC, Metro area. He has edited several books and written articles on materiality, cellphones and Oceania's cultural and natural heritages.

Tilly Blyth is a professor of museum studies at the University of Leicester where she brings together sector knowledge, with creative practice and academic research, to consider the relevancy and structural challenges of the sector. In her previous role as head of collections and principal curator at the Science Museum in London, she was responsible for a curatorial, research and library and archive teams that developed award-winning galleries and exhibitions on subjects as diverse as Mathematics, Robots, Cosmonauts: the Russian Space Story and Illuminating India: 500 Years of Science and Technology. Tilly was lead curator on the Information Age gallery, looking at the invention and use of information and communication technologies through a series of transformative narratives. Sheco presented a landmark 20-part radio series for the BBC on the art of innovation, including a book and an exhibition that looked at the shared cultures of art and science over the past 250 years. Tilly is a member of the British Academy of Film, Television and the Arts (BAFTA); a fellow of the Royal Society for Arts, Manufactures and Commerce (RSA); and a trustee of the Bletchley Park Trust and the Raspberry Pi Foundation, which aims to put the power of digital making into the hands of people all over the world.

Rachel Boon is an experienced museum professional with a background in history of science, technology and medicine, and public engagement. She is Curator of Technology and Engineering at the Science Museum where she is responsible for the Computing and Data Processing collection. Rachel has worked on several exhibitions at the Science Museum, including The Art of Innovation (2019–2020) and Churchill's Scientists (2015). Previously, she was an exhibitions and interpretation officer at the Imperial War Museum and has been part of the learning teams at the Science and Industry Manchester and Science Museum. Her research interests include industrial research and development, histories of infrastructures, and British scientific establishments. She completed her PhD in 2020 on the history of the Post Office Research Station, Dollis Hill.

David C. Brock is a historian of technology, curator, and director of CHM's Software History Center. He focuses on software history, as well as oral history, and is a key member of the Museum's content team. Brock's publications include *Thackray, Brock and Jones, Moore's Law: The Life of Gordon Moore, Silicon Valley's Quiet Revolutionary* (Basic Books, 2015); *Lécuyer and Brock, Makers of the Microchip: A Documentary History of Fairchild Semiconductor* (MIT Press, 2010); and *Brock (ed.) Understanding Moore's Law* (CHF, 2005). He has contributed to films, television productions, broadcast events, and museum exhibits. Brock is on Twitter as @dcbrock.

Martin Campbell-Kelly is Emeritus Professor in the Department of Computer Science at the University of Warwick. His books include *Computer: A History of the Information Machine*, 4th edition, 2023 (co-authored with William Aspray, Jeff Yost, Honghong Tinn and Gerardo Con Diaz) and *From Airline Reservations to Sonic the Hedgehog: A History of the Software Industry*, 2003. Professor Campbell-Kelly is a fellow of the British Computer Society, a fellow of the Learned Society of Wales, and a former trustee of the National Museum of Computing, UK. He is a member of the editorial boards of the *IEEE Annals of the History of Computing*, the *International Journal for the History of Engineering and Technology*, and founding editor of the Springer Series in the History of Computing.

Simona Casonato is Curator of Media, ICT, and Digital Culture collections at the Museo Nazionale Scienza e Tecnologia Leonardo da Vinci, Milan, Italy. She has been working at the museum since 2003 as a multidisciplinary specialist, coordinating the production of documentary videos, carrying on historical research on museum collections, and curating public events. Her interests include the social history of technology and media, heritage studies, oral history, and ethnographic methodologies. She collaborates with Italian and European universities for research and teaching. In 2022 and 2023 she lectured at the Design Department of Politecnico di Milano.

Corinna Gardner is a senior curator of design and digital at the V&A. Corinna leads the museum's Rapid Response Collecting Program and her research focuses on product and digital design and their role in public life. Recent projects include a new gallery for twentieth and twenty-first century design at the V&A in London, the development of a digital design collecting strategy for the museum, and the international touring exhibition *Plastic: Remaking Our World* in collaboration with the Vitra Design Museum.

Hansen Hsu is a historian and sociologist of technology and curator at the CHM Software History Center. He works at the intersection of the histories of personal computing, graphical user interfaces, object-oriented programming, and software engineering. After receiving his BS in EECS from UC Berkeley in 1999, Hsu joined Apple through 2005 as a quality assurance engineer in the AppKit framework group. Hsu received his MA in history from Stony Brook University in 2007, and his PhD in science and technology studies from Cornell University in 2015, with a dissertation exploring the culture of the Apple software developer community.

Natalie Kane is a curator of digital design at the V&A in London, where they are responsible for the care, research, and collection of the museum's digital design collection. Most recently, Natalie curated the official UK entry to the Milan Triennale in 2019 with the work of research agency Forensic Architecture. In 2019, they received the Art Fund New Collecting Award to grow the museum's collection of digital design. Natalie was the principal investigator of the AHRC-funded 'Towards a National Collection' Foundation Project 'Preserving and Sharing Born Digital and Hybrid Objects,' which aimed to address challenges of collecting digital objects in museums and memory institutions. They are one-half of curatorial research project Haunted Machines and sits on the advisory board of the Society of Computers and Law.

Kimon Keramidas is currently Head of Digital Content and Strategy at the Rubin Museum of Art in New York City, which promotes understanding of the ideas, cultures, and art of Himalayan regions. Prior to the Rubin, Kimon was an associate professor of experimental humanities and social engagement at NYU, assistant professor and director of the Digital Media Lab at the Bard Graduate Center (BGC), and co-director of the International DH Research Center at ITMO University in St. Petersburg, Russia. Other collaborations include work with the Smithsonian National Museum of Asian Art (formerly the Freer|Sackler), the State Hermitage in St. Petersburg, the University of Leicester, and NYU Abu Dhabi and Zayed University in the United Arab Emirates. Recent projects include the digital portion of *Project Himalayan Art* (https://projecthimalayanart. rubinmuseum.org/), the Rubin's initiative to support the incorporation of Himalayan art and cultures into humanities teaching curricula; co-curating *The Sogdians: Influencers on the Silk Roads* (https://sogdians.si.edu/), a digital global art history project at the National Asian Art Museum; and curating *The Interface Experience: Forty Years of Personal Computing*, a transmediated exhibition/book/digital experience on the history of personal computing at the BGC (http://bit.ly/The-Interface-Experience, http://interface-experience.org/). Kimon is also co-founder of The Journal of Interactive Technology and Pedagogy and New York City Digital Humanities (NYCDH).

Andrea Lipps is Head of the Digital Curatorial Department and Associate Curator of Contemporary Design at Cooper Hewitt, Smithsonian Design Museum in New York. She leads the museum's efforts to collect and care for new media types in its permanent collection and innovates scholarship in the field. She also co-chairs the museum's Responsive Collecting Initiative. Lipps further conceives and organizes ambitious, award-winning exhibitions, including *An Atlas of Es Devlin* (2023), *Nature – Cooper Hewitt Design Triennial* (2019), *The Senses: Design Beyond Vision* (2018), and *Beauty – Cooper Hewitt Design Triennial* (2016). An accomplished writer and editor, she authors and edits publications, essays, and scholarly articles on contemporary design and digital collecting. Lipps is a regular visiting critic, lecturer, and thesis advisor; participates on international design juries; and frequently moderates and speaks at events, symposia, and academic conferences on contemporary design and curatorial practice.

Lisa McGerty is currently Chief Executive Officer at the Centre for Computing History in Cambridge, England. She was one of the Centre's founders. She is currently leading on projects about the 1950s–1960s Lyons Electronic Office (LEO) Computers and about the under-acknowledged women in computing history. She is also involved in projects on the role museums can play in both computing education in the UK, particularly in the early years, and in the participation of all genders in IT-related learning and careers. Her research interests more generally focus on the people behind the first-generation computers; the intersections of genders, technologies, and domestic spaces; and the impact of users on narratives around computing history. She is always seeking the hidden stories in computing history.

Juhee Park is a digital museum researcher, lecturer, and curator with over 15 years' experience working in cultural and heritage sectors in the UK and South Korea. She is currently a senior manager at the Hyundai Motor Company, Republic of Korea, where she manages a museum building project. Previously, she was a curator at the Gyeonggi Children's Museum, a research fellow at the V&A, and a senior researcher at the KI-ITC Augmented Reality Research Center, Korea Advanced Institute of Science and Technology (KAIST). Her research interests include digital heritage, virtual exhibitions, digital collecting, and the ecology of AI museums. She holds a BE in computer engineering from Korea University, an MS in culture technology from KAIST, and a PhD in museum studies from University College London.

Mark Priestley is a research fellow at The National Museum of Computing, Bletchley Park, UK. He worked as a programmer and then university lecturer in software engineering before focusing his research activities on the history and philosophy of computing. His primary focus is on the development of the theory and practice of programming, and he undertook a PhD studying the influence of logic on the early development of programming languages. From 2012 to 2016 he was engaged in the *ENIAC in Action* project (MIT Press, 2016), analyzing and reconstructing archival programming material related to ENIAC. Recent projects have included studies of the operation and use of the Colossus machines used at Bletchley Park in World War II and of John von Neumann's programming work in the 1940s. The results of the latter are described in his latest book, *Routines of Substitution: John von Neumann's Work on Software Development, 1945–1948* (Springer, 2018).

Lara Ratnaraja is an independent cultural consultant who specializes in culture and diversity, innovation, leadership, collaboration, and cultural policy and placemaking within the cultural, HE, and digital sectors. She also co-produces a series of cultural leadership programs for people from diverse backgrounds linked to geographical places and curates a digital Conference called Hello Culture. She has developed, managed, and delivered successful ERDF, ESF, ACE, AHRC bids, and programs working within the public, HE, and commercial sectors and was director for the creative, cultural, and digital industries at Business Link WM from 2003 to 2011. She has significant experience in both the application and implementation of cultural policy and strategy and the public sector framework within which these operate. Lara is on the board of Derby Theatre, Vivid

Projects, and Coventry Biennial and the Advisory Group for SHOUT Festival. She is on the UK Council for Creative UK and the Equality Monitoring Group for Arts Council Wales.

Dag Spicer is an engineer and historian by training and acts as senior curator at the Computer History Museum (CHM). He is responsible for the growth, exhibition, and interpretation of the museum's physical object collection, the largest collection of computing-related objects in the world. Since beginning at CHM in 1996, Dag and his curatorial colleagues have shaped CHM into a world-class resource for the study of the history of computing and its complex impacts on humanity.

Mai Sugimoto is a professor at the Faculty of Sociology, Kansai University, in Japan. She works on the histories of computer science and artificial intelligence from the mid-twentieth century onward. Her works include *"Jinko Chino" Zenya* (Artificial Intelligence Eve), 2018 (in Japanese). Sugimoto received her PhD in philosophy and history of science from Kyoto University in 2013. She is a member of the editorial board of *Historia Scientiarum, The International Journal of the History of Science Society in Japan*, and an editorial board member of Johns Hopkins University Press's Studies in Computing and Culture Series.

Marc Weber is curatorial director of the Internet History Program (computerhistory.org/nethistory) at the Computer History Museum (CHM). He launched web history as a topic starting in 1995, with crucial help from Sir Tim Berners-Lee and other pioneers. He co-founded two of the first organizations in the field and has curated over a dozen galleries or exhibits at CHM on networking, mobile, and AI. He presents and publishes widely and consults to media, patent firms, and filmmakers. He serves on the editorial board of *Internet Histories: Digital Technology, Culture and Society*. Weber has taught at the University of California and holds bachelor's degrees in neurobiology and in creative writing from Brown University.

Acknowledgments

Research leading to the publication of this book was conducted as part of the Circuits of Practice Project, thanks to a Research Grant awarded by the Arts and Humanities Research Council (AHRC) (AH/T00276X/1, PI Simone Natale). The project explored the construction and dissemination of narratives about histories of computing within museum environments. Our deepest thanks and gratitude go to our partners, which included Bletchley Park, BT Archives, the Centre for Computing History, the National Science and Media Museum, the Science Museum and the Victoria & Albert Museum in the UK, the Computing History Museum in the US, the Museo Nazionale Scienza e Tecnologia Leonardo da Vinci in Italy, and the National Museum of Emerging Science and Innovation 'Miraikan' in Japan. We would like to thank, moreover, the project's scientific advisors, Joshua Bell, Kimon Keramidas, Maholo Uchida, Mar Hicks, Sabina Mihelj, and Lara Ratnaraja, for their feedback that helped improve and validate the project's findings.

The project took place during a moment of pandemic, to which we were forced to adapt. We would like to thank our partners, as well as the AHRC, Loughborough University, and the University of Leicester, for their flexibility and support, without which we would have simply not been able to run the project. We thank, moreover, all people who contributed with enthusiasm and generosity to this collective endeavor even as the effects of the pandemic moved most of our activities online.

The Circuits of Practice Book Club, led by Petrina Foti, began as a research output but soon grew into something much larger and more enduring. The Book Club has proved to be a valuable way to connect with our project partners and engage with academic literature outside of our individual professional interests, fostering an interdisciplinary community of academic peers. We would like to thank all Book Club participants, past and present, for their generosity in sharing their time, their expertise, and their insights.

A further initiative promoted by Circuits of Practice was the Digital Atelier, an ensemble developing alternative, challenging, creative ways of understanding and sharing the project's findings, under the lead of Kimon Keramidas and Ross Parry. Our thanks go to the three researchers and designers – Elisabetta

Gomellino, Molly Shand, and Amelia Taylor – who animated the Atelier, moving from the project's findings to generate new provocations and insights, each unsettling as much as demonstrating our new understanding of computing history in the museum.

The Centre for Research in Communication and Culture (CRCC) at Loughborough University provided seed funding that was instrumental to conduct preliminary research and to shape the key ideas for the project. Thanks to this funding, we organized a networking event titled *Constructing Histories of Computing and Digital Media in Museum Environments*, which involved many of the future partners for the project. Our thanks to CRCC and the Department of Social Sciences at Loughborough University, as well as to Thais Sarda and Yingzi Wang for contributing to the event's organization.

The Institute for Digital Culture at the University of Leicester generously provided funding and support to fund editorial activities that helped prepare the manuscript for submission. Thank you to Tom Eaton for helping with editing and proofreading, as well as our Editor at Routledge, Heidi Lowther, and Editorial Assistant Heeranshi Sharma for providing feedback and assistance.

Across several years, there are many people who offered feedback and who participated in some form to the project's activities. Our deepest thanks go to Geoffrey Belknap, Paolo Bory, Kathryn Brown, Elizabeth Bruton, Iliana Depounti, James Elder, Patricia Falcao, Elinor Groom, Lise Jaillant, Francesca Olivini, Sarah Ledjmi, Phillip Roberts, Laura Ronzon, Jeremy Thackray, and Jochen Viehoff, and to many other who might not be mentioned here.

This book is dedicated to the memory of Silvio Henin, a life in computing history.

Introduction

Museums and the history of computing

Simone Natale

Digital technologies are a ubiquitous presence for billions of people around the world, shaping not only their everyday experiences but also their social lives, their access to information, and their identity (Papacharissi, 2002). As part of this process, narratives, discourses, and imaginaries about these technologies have acquired an increasingly prominent role in the public sphere. News media outlets constantly report and discuss events and issues relating to digitalization, artificial intelligence, and social media platforms; mainstream films and television series relay stories about key figures and events within the 'information revolution'; protagonists in computing history such as Alan Turing and Steve Jobs are celebrated as some of the most significant figures of the twentieth- and twenty-first centuries.

It is evident that we are in the process of building and defining a new range of narratives about the emergence, development and future of computing and digital media (Natale, 2016). These narratives are important for three key reasons: firstly, they help us to understand how digital media came to play such a key role in our lives and to comprehend the new directions and accompanying challenges that digital media present for society and everyday lives (Balbi & Magaudda, 2018). Secondly, narratives have practical and material effects: as a wealth of research has demonstrated (Mansell, 2012; Mager & Katzenbach, 2021), narratives about digital technologies inform public debates that in turn shape the governance of these technologies. Thirdly, such narratives help to shape our understanding of who we are: scholars in memory studies show how cultural heritage and collective memory, as mediated through digital technologies, contribute in a crucial way to shaping people's identities and their understandings of the world (Graham & Howard, 2012).

Despite a growing focus upon these narratives and the complex social imaginaries revolving around digital media – see for example, amongst others, Bory (2020) and Ernst and Schröter (2021) – the role of museums in constructing these narratives has been largely disregarded.[1] Although the topic has started to receive attention within the museum studies field (see, for example, Weber, 2016), most studies about narratives and imaginaries relating to computing and the digital have continued to rely on sources such as news

DOI: 10.4324/9781003424703-1

media and popular publications, with little or no attention directed towards the museum space. This is surprising, considering how both permanent and temporary exhibitions in museums have been active in sharing information, reflections, and compelling stories about digital technologies to large audiences around the world. As part of a burgeoning interest in the digital, the last few years have seen numerous new museums dedicated to histories of computing and digital media established in many countries around the world. Moreover, existing institutions have made greater efforts to integrate histories of computing into their exhibitions. But this lack of attention towards museums is even more surprising if we consider the distinctive contribution museums bring to the construction of a new cultural heritage about the digital. By presenting nuanced, dialogic, interactive stories and experiences of technological change, museums empower audiences and users by providing them with a toolbox to understand, interpret, and question the trajectories and implications of technological change and the role of the digital in shaping societies (Blyth, 2013; Burton, 2013; Cameron, 2007).

This book examines how museums shape digital heritage, that is to say, the cultural heritage surrounding computer technology and digital media. While both theoretical and practice-based explorations of digital technology in the museum sector have largely interrogated digital media as tools for museum practice, less attention has been given to the subject of digital technologies as 'materials' for museums to collect, exhibit, and preserve for the future. *Museums and Digital Histories* contributes to address this gap, seeking to show how and why museums are playing a crucial role in preserving the rich heritage of digital technologies and in making it relevant to the public.

Cultural heritage institutions have collected materials relevant to the history of computing and digital technologies for a number of decades. However, decisions about what must be conserved, and the approaches taken to achieve this, are far from established institutional principles and codes of practice. As Marc Weber (2016) has pointed out, choices made by cultural heritage institutions today are closely related to the historical narratives that are produced and privileged by different institutions and groups in different national and cultural contexts. Such narratives shape decisions about what counts as historically significant and what does not. This in turn influences decisions about what counts as part of the historical record. Reflecting on the dynamics and practices through which histories of digital technologies are constructed and disseminated in museums is therefore an important issue with far-reaching implications for ongoing debates about conservation and exhibition practices in the museum.

Narratives and artefacts relating to the history of computing pose significant challenges to museums. By examining and reflecting on these challenges, this book sheds light upon how existing curatorial practices adapt to new subjects and how innovative curatorial practices can be developed to tackle new kinds of problems and questions. Firstly, strategies adopted for the preservation and exhibition of computing artefacts need to adapt to a variety of hardware and software objects and to changing technological standards and supports (Foti, 2018). Secondly, the complexity of technological and social change makes constructing clear narratives about computing particularly arduous (Blyth, 2013). Thirdly, actors informing technological change are manifold, including users, developers, companies, and states, and their agency is often complex and hard to capture and demonstrate (Keramidas, 2015). Fourthly, and perhaps most centrally for this book, narratives about technological change and digital media never exist in a void. As museum curators choose which trajectories to highlight and present to visitors, they have at their disposal a broad range of perspectives upon technology (such as technological determinism, for example) and work within a discursive space that is already saturated by narratives such as representations of individual 'heroes' or the stories circulated by public relation teams within the tech industry (Streeter, 2015; Natale et al., 2019).

Indeed, it is not only digital media that are constantly subject to change. The terms in which media are talked about are also in endless transformation. Consider, for example, how a word such as 'virtual' has entered in and out of fashion across the last three or four decades. Consider also how a single corporation such as Meta recently brought to the very core of public debate the concept of the 'metaverse' and how this very same concept already started to lose credibility as it became clear that the company's strategy and execution failed to fulfil the bold ambitions that the term embodied (Roquet, 2023). As narratives about histories of computing circulate in the public domain, museums are invited to incorporate and respond to new directions that characterize the representation of computing and digital media in the broader public sphere. They provide a key space to historicize not only the technology but also the concepts and the narratives that accompany them (Balbi et al., 2021) and, at the same time, serve to document the wider transformations of the ways we think, talk about, and use digital technologies.

Bringing digital media to life

As the contributions collected in this volume show, studying how histories of digital media and computing are integrated within museum environments provides a distinctive and essential entry point into digital heritage. Museums not only tell stories and explain digital technologies, they invite and stimulate audiences and users to look for new ways to think about them and to therefore consider new ways to interact with and use media that have become a central presence – at times an oppressive presence, as Tero Karppi (2018) emphasizes – in all facets of our lives, across communication, education, work, entertainment, social activities, and sex. Approaches to digital literacy have highlighted how the ability to engage with digital technologies in a competent and reflexive way can only be fostered through a deep engagement with the technologies themselves (McDougall et al., 2018). Given that both multimedia and interactive design features are typical of the contemporary museum (Parry, 2010), exhibition spaces are ideal environments to pursue this agenda. Therefore, museums are not just one of the actors that create narratives and discourses about digital media, they have the potential to transform the very nature of these narratives, as users are brought to the very center of the stories and trajectories presented in the exhibitions.

In this sense, it is unsurprising that the notion of the *lives* of digital objects has become central to the research presented in this collection. Rosemary Joyce and Susan Gillespie (2015) have criticized the use of the 'life' metaphor to consider material objects, contending that such a notion replicates a cultural bias that projects the dynamics of human lives onto objects. We contend, however, the idea that things have 'lives' is useful precisely insofar as it helps to illuminate how such projections and appropriations shape the ways material objects are appropriated and narrated across time. Things are inseparable from the social and cultural values that people attach and project onto them (Appadurai, 1986). People, institutions, and social groups make objects 'alive' by projecting uses, perceptions, narratives, and representations onto them (Natale, 2016). Thus, to follow the trajectories of digital objects – indeed, any kind of object – in a museum environment entails revealing the trajectories that shape our perception and engagement with them. The concept of 'lives,' in this sense, accounts for the changing meanings and positions that objects assume across time, and to the close interrelationship that exists between the trajectory of objects and the experiences of people who interact and project sense onto them. These include visitors, curators, practitioners, volunteers, and other individuals and groups that animate objects and stories about digital media through their contribution and agency in and beyond museum spaces.

The perspective of museum-based researchers, largely represented in this book, provides an entry point into the relational nature of digital heritage and how histories of computing and the objects collected in the museum take up new lives as they become entangled with the work and lives of museum practitioners. The museum functions as a laboratory in which meanings, uses, and definitions of digital media are self-reflectively negotiated and where the relational circumstances that characterize digital objects becomes manifest. An important role, in this context, is played by the productive encounters between the material character of objects related to the histories of computing, the particular character and institutional culture of each museum, and the activation of social meanings and affect that inform the experiences of curators and practitioners within the museum collection (Geoghegan & Hess, 2015). As they become part of the exhibition, digital media continue to mobilize and produce affective value through a triangle of affect between the curators, the objects and exhibitions, and the visitors.

Moreover, the relational character of digital objects emerging in museum practice resonates with the multiple ways in which visitors navigate exhibitions, engage with digital objects like those they encounter in their everyday life, and project their own previous experience and ideas about digital media. As digital media become repositories of multiple social uses, meanings, and exchanges when they are created, used, circulated, and eventually discarded within everyday life, they also establish a relational and iterative social 'circuit of meaning' within the museum environment. Rather than diverging from the characteristics of digital objects outside of the museum, the trajectories of digital objects in museum collections and exhibitions prove in this sense to be in a relationship of continuity with the material, social, and narrative trajectories of digital objects outside the museum.

The book

Bringing together leading museum and university researchers and mobilizing cross-cutting debates and approaches in areas including museum studies, cultural heritage, the history of technology, anthropology, and media studies, this book aims to challenge researchers, students, and practitioners to think critically about what 'digital' is when examined not only as a tool but as a cultural object within the museum. The book is organized into four sections, each of which engages in a different way with the metaphor of the 'lives' of digital media.

Part I, 'Lives Narrated Through Computer History,' examines the capacity of museums to bring to the surface the experiences of users, technologists, and other people who have contributed to shaping histories of computing technologies. As many have argued, objects have social lives of their own. However, these lives are always entangled with the lives of the people who use, own, or interact with them. The chapters collected in this first section therefore illuminate how hardware and software artefacts in museum collections function as a prism through which to understand the 'lives' of the objects as well as of their users. In Chapter 1, Joshua Bell reflects on his experience as curator of a major exhibition about cellphones at the National Museum of Natural History (NMNH) in Washington, DC. Bringing together perspectives from anthropology, natural history, and people's everyday experiences, the stories and objects displayed in the exhibition reveal how technology is a central part of humanity and is thus situated within what the NMNH labels as nature. In Chapter 2, Simona Casonato gives an account of the biographies of specific objects in storage at the Museo Nazionale Scienza e Tecnologia Leonardo da Vinci in Milan, Italy. Far from being discarded items that remain silent in the museum's storage spaces, Casonato shows how these objects are constantly revived as their meanings emerge as part of the complex entanglements between the circumstances of their acquisition, the agency and affect of the museum's curators and practitioners, the visitors and users who encounter them during the museum's guided tours, and the social lives of the objects themselves.

Part II of the book, 'The Life Inscribed on Computer Technology,' considers the role of hardware and software artefacts to narrate histories of modern computing within museums. It illuminates the material, social, and discursive dynamics through which digital objects become lively repositories of evidence, narrative, and affect in museums' collections and exhibitions. In Chapter 3, Martin Campbell-Kelly and Mark Priestley consider different ways to prepare artefacts related to the histories of computing and the digital for exhibition and preservation purposes. Focusing on the case of The National Museum of Computing at Bletchley Park, UK, their essay sheds light on a veritable 'community of machines' where diverse curatorial practices preserve material and technical elements that would otherwise be lost from the historical record. In Chapter 4, Natalie Kane, Corinna Gardner, and Juhee Park explore as a case study the Victoria & Albert Museum's acquisition of the social media platform WeChat. The many challenges this entailed and the responses that the curators developed to address these challenges help to reveal the process through which museum institutions develop new protocols and approaches to incorporate new kinds of objects, such as online-based software, into museum spaces.

Part III of the book, 'Living Computing History Collections,' frames the history of computer technology in terms of the museum collection. It examines how the museum has responded to the challenge of the intangible aspects of computer technology such as digital materials and data itself. While it is possible for a museum to collect and exhibit digital objects, the process of archiving does not necessarily encompass digital technology's associated ecosystem. In Chapter 5, Rachel Boon and Tilly Blyth illustrate the shift from a modality of a collection based on a narrative of heroes, which tends to present scientific innovation through a select range of celebrity machines and key human protagonists, to a perspective and practice that focuses on material culture as a lens through which to explore how digital technologies and data are culturally, socially and politically constructed. In Chapter 6, Petrina Foti asks how deep changes in the essence of computing can impact on how museums narrate histories and the present of digital technologies. Taking up a momentous challenge such as the need to tell the story of the technology used in the making of quantum computing, the chapter reflects on museums' need to constantly interrogate meaning-making and human culture. It shows that making sense of something such as quantum computing is not just one element of museums' work but an ideal example of the challenges and opportunities that shape their mission.

Part IV, 'Lived Practice with Computing History,' turns to the institutional context in which computing histories are always inevitably made. As the objects of modern computing in the museum become evidence of multiple social, technical and personal narratives, they also enter a unique organizational context. Consequently, exhibitions on museum computing are also manifestations of the museum's own identity and relationship with digital technology. In Chapter 7, David Brock, Marc Weber, Dag Spicer, and Hansen Hsu consider the evolution of the digital heritage collection at the Computer History Museum in California. As the institution developed frameworks and protocols to collect and exhibit software, a 'stack' for contending with digital heritage emerged within the museum, operating on different timescales, from permanent collection to ephemeral events, whilst entering into a relationship with different communities and craft practices. In Chapter 8, Lisa McGerty examines the case of the Centre for Computing History in Cambridge, UK. McGerty shows how the 'actuality of the museum,' that is, the organizational context within which CCH is situated, impacts the histories of computing that are present, performed, and produced in the museum environment. Through the lens of two very different organizational contexts, the two contributions in this section of the book therefore explore distinctive ways through which computing history evolves within the 'lived' institutional environment of the museum.

The book also contains four shorter contributions from Mai Sugimoto, Andrea Lipps, Kimon Keramidas, and Lara Ratnaraja. We call these sections "Provocations," and they bring to the fore some of the threads, inspirations, criticisms, and issues that this book could only just touch upon. Indeed, one of the most evident results of the research that has been developed in this project is an acknowledgment of the plural character of narratives about the history of computing and digital media. There is no single way to talk about digital technologies, and even about any single digital technology, such as, for instance, the smartphone or the combination of software that powers voice assistants such as Alexa or Siri. The telling of these stories is always situated in specific cultural, social, national, political, economic, linguistic, and institutional contexts, which deeply inform which narratives are told about the digital heritage. Cases examined in this book are mostly based at sites within the northern hemisphere, but we hope that our reflections on their specificity and on the importance of context can work as an invitation for the sharing of diverse narratives and trajectories in museums from around the world.

The book emerged from research in the AHRC-funded project 'Circuits of Practice: Narrating Computing Histories in Museum Environments.' Running between 2019 and 2021, the project brought together researchers based at both universities and museums to interrogate how museums narrate modern computing (see Natale et al., 2022). In each individual chapter within this book, the reader will be able to find traces of the common preoccupations, ideas, and excitements that developed throughout two years of research, collaborations, and discussions and that make this book a truly collective endeavor. In the same way that within an electronic circuit there exist electrical connections between diverse components, which enable complex operations to be performed, the 'Circuits of Practice' within this book have established intellectual, practical, and affective connections between a community of researchers and practitioners. This has enabled the production of new knowledge and ideas in a way that could never be achieved without the sum of its components.

Note

1 A working definition of museum is provided by ICOM: "A museum is a not-for-profit, permanent institution in the service of society that researches, collects, conserves, interprets and exhibits tangible and intangible heritage. Open to the public, accessible and inclusive, museums foster diversity and sustainability. They operate and communicate ethically, professionally and with the participation of communities, offering varied experiences for education, enjoyment, reflection and knowledge sharing" (ICOM, 2024).

Reference List

Appadurai, A. (1986). *The social life of things: Commodities in cultural perspective*. Cambridge University Press.

Balbi, G., & Magaudda, P. (2018). *A history of digital media: An intermedial and global perspective*. Routledge.

Balbi, G., Ribeiro, N., Schafer, V., & Schwarzenegger, C. (2021). *Digital roots: Historicizing media and communication concepts of the digital age*. De Gruyter.

Blyth, T. (2013). Narratives in the history of computing: Constructing the information age gallery at the science museum. In A. Tatnall, T. Blyth, & R. Johnson (Eds.), *Making the history of computing relevant* (pp. 25–34). Springer.

Bory, P. (2020). *The internet myth: From the internet imaginary to network ideologies*. University of Westminster Press.

Burton, C. P. (2013). The teenage 'baby' on show. In A. Tatnall, T. Blyth, & R. Johnson (Eds.), *Making the history of computing relevant* (pp. 274–284). Springer.

Cameron, F. (2007). Beyond the cult of the replicant: Museums and historical digital objects: Traditional concerns, new discourses. In F. Cameron & S. Kenderdine (Eds.), *Theorizing digital cultural heritage: A critical discourse* (pp. 49–71). MIT Press.

Ernst, C., & Schröter, J. (2021). *Media futures: Theory and aesthetics*. Springer.

Foti, P. (2018). *Collecting and exhibiting computer-based technology: Expert curation at the museums of the Smithsonian Institution*. Routledge.

Geoghegan, H., & Hess, A. (2015). Object-love at the science museum: Cultural geographies of museum storerooms. *Cultural Geographies, 22*(3), 445–465. https://doi.org/10.1177/1474474014539247

Graham, B., & Howard, P. (2012). *The ashgate research companion to heritage and identity*. Ashgate.

ICOM. (2024). *Museum definition*. https://icom.museum/en/resources/standards-guidelines/museum-definition

Joyce, R. A., & Gillespie, S. D. (2015). *Things in motion: Object itineraries in anthropological practice*. SAR Press.

Karppi, T. (2018). *Disconnect: Facebook's affective bonds*. University of Minnesota Press.

Keramidas, K. (2015). *The interface experience: A user's guide*. Bard Graduate Center.

Mager, A., & Katzenbach, C. (2021). Future imaginaries in the making and governing of digital technology: Multiple, contested, commodified. *New Media & Society, 23*(2), 223–236. https://doi.org/10.1177/1461444820929321

Mansell, R. (2012). *Imagining the internet: Communication, innovation, and governance*. Oxford University Press.

McDougall, J., Readman, M., & Wilkinson, P. (2018). The uses of (digital) literacy. *Learning, Media and Technology, 43*(3), 263–279. https://doi.org/10.1080/17439884.2018.1462206

Natale, S. (2016). Unveiling the biographies of media: On the role of narratives, anecdotes and storytelling in the construction of new media's histories. *Communication Theory, 26*(4), 431–449. https://doi.org/10.1111.comt.12099

Natale, S., Bory, P., & Balbi, G. (2019). The rise of corporational determinism: Digital media corporations and narratives of media change. *Critical Studies in Media Communication, 36*(4), 323–338. https://doi.org/10.1080/15295036.2019.1632469

Natale, S., Parry, R., & Foti, P. (2022). *Circuits of practice research report: Narrating histories of computing and digital media in museum environments*. School of Social Sciences and Humanities, Loughborough University.

Papacharissi, Z. (2002). The virtual sphere: The internet as a public sphere. *New Media & Society, 4*(1), 9–27. https://doi.org/10.1177/146144402222262

Parry, R. (2010). *Museums in a digital age*. Routledge.

Roquet, P. (2023). Japan's retreat to the metaverse. *Media, Culture & Society, 45*(7), 1501–1510. https://doi.org/10.1177/01634437231182001

Streeter, T. (2015). Steve Jobs, romantic individualism, and the desire for good capitalism. *International Journal of Communication, 9*, 3106–3124. https://ijoc.org/index.php/ijoc/article/view/4062/1473

Weber, M. (2016). Self-fulfilling history: How narrative shapes preservation of the online world. *Information & Culture, 51*(1), 54–80. https://doi.org/10.7560/IC51103

Part I

Lives narrated through computer history

1 Unseen connections

Exhibiting the global stories of cellular telephony at the Smithsonian's National Museum of Natural History

Joshua Bell

What is globalization? How does it shape communities? What are paradigmatic objects within this process? While these questions have been an implicit aspect of my academic career, they took on new relevance when I took up my current position as curator of globalization at the Smithsonian's National Museum of Natural History (NMNH) in 2008. Part of a repositioning of anthropology at NMNH to address more thematic research and curatorial concerns, this title was meant to signal the ways in which anthropology in museums, despite long-term histories and ingrained problems, is very much responding to the challenges and expectations of the twenty-first century. The title was a provocation for me to rethink the global dimensions of my research in the Purari Delta of Papua New Guinea and which stories could be told about the histories and relations materialized in NMNH's collections and displays. It was also an invitation to develop interdisciplinary projects that bring the diverse processes we term globalization into view in novel ways (Bell, 2017). In what follows, I give an account of some of these transformations and realignments as they relate to an exhibit that I curated for NMNH – *Cellphone: Unseen Connections* and which opened in June 2023.

Histories

With the advent of the iPhone and the popularization of smartphones in the late 2000s, I became intrigued with the global supply chains that are critical to the manufacture of this device, the places from which these materials are extracted, the people that are critical to the supply of materials, and what the ecological implications of when we are done with these objects (Bell et al., 2018a; Kuipers & Bell, 2018). In 2011 I began collaborating with linguistic anthropologist Joel Kuipers of George Washington University, leading a research team of graduate and undergraduate students to examine the global relations that are materialized in cellphones. Conducting an ethnography on third-party repair technicians in Washington, DC (Bell et al., 2018b; Kuipers

DOI: 10.4324/9781003424703-3

et al., 2018), we then began working with linguistic anthropologist Alexander Dent, also of George Washington University, to focus on the social, material, and linguistic aspects of cellphone use amongst students in two high schools in Washington, DC (Bell et al., 2021; Dent et al., 2022). This work became a means by which to understand how this technology is transforming social experience.[1]

Knowing how difficult temporary exhibitions at NMNH can be, I had conceived of this exhibit's development as synergistically connected to the research outlined previously but not bound to it. In 2011 I convened scholars from the fields of anthropology, design, environmental science, arts, geology, linguistics, and public health, to talk about what an exhibition about the cellphone could look like.[2] The workshop focused on the interconnected material, linguistic, and design aspects of the cellphone.[3] Fundraising demands for the new iteration of NMNH's Dinosaur Hall took precedence, and the development of the exhibition was put on pause. In the interim, I carried out public programs and worked with interns to further develop the exhibit. Eventually in 2016 the exhibition was approved, and a core development was formed, consisting of a project manager/exhibit designer, writers, designers, media consultants, an educator, a curator (myself), and exhibition/research assistants. Alongside of this, we formed an advisory team of academics, as well as industry and museum professionals. The development of the exhibition went through five phases – concept phase, design phase, and refined design phase – each requiring approval from the museum leadership to move forward. The challenges posed by the COVID-19 pandemic meant remote meetings, creative collecting practices, and a new timeline for the opening date of the exhibition. In 2021 NMNH received a gift of three million dollars from Qualcomm, the main sponsors of the exhibition, alongside a gift of one million dollars from T-Mobile. These gifts covered the costs of the exhibit fabrication and installation and are allowing us to develop a suite of educational and outreach materials.

From the beginning the exhibition was intended for the demographic of 12–24-year-olds. While the show will be of interest to all ages, we focused on this specific demographic due to two key factors: firstly, within the United States the average teenager acquires their first smartphone at 12 years old, and secondly, if and when young adults lose interest in museums and science, this is the age at which this tends to happen (Anderhag et al., 2016; Lei et al., 2019).[4] This is also a demographic that has grown up in a world immersed in cellular telephony and touch screens. The exhibition offers a unique opportunity to teach science through a contentious device that is a source of mixed commentary and lively discussion (Dent et al., 2022). We organized the exhibit through a compelling set of stories presented in an informative and yet jargon-free and conversational style. Throughout the design process, we engaged young adults through focus groups to discuss the exhibition's content, themes, and messages. We used their feedback to inform and reformulate our design and interpretation strategy.

Unseen connections: the exhibition

A central goal of the exhibition is to convey that technology forms a central part of human experience, and that this relationship is situated within NMNH's remit of 'nature.'[5] The exhibit's relational perspective draws inspiration on the one hand from Indigenous Oceanic cultural sensibility (Bell & Geismar, 2009) and on the other from the perspective of the interrelated nature of humanity espoused by Martin Luther King Jr. (King, 2010). Curatorially, I was also inspired by similar sentiments articulated by Jane Bennett (2010), who has emphasized the more-than-human as part of the push to rethink of ecological obligations, and by Native scholar Kim Tall-Bear's (2019) call to engage with the more inclusive Indigenous notions of kin and kin-making. Engaging this confluence of ideas in a study of plastic pollution informed by an Indigenous framework, Max Liboiron (2021) has articulated how this set of frameworks does and does not work with electronics and human-made pollutants. Liboiron reminds scholars about the danger of appropriating Indigenous concepts and challenges scholars to see the relationality without dispossessing others. Mindful of this criticism, my hope is that if nothing else, this exhibition becomes a means by which visitors see their devices anew, begin to ask questions about their wants and needs, and that they think about how their consumption ties them into wider local and global networks (Lepawsky, 2018). While this is a particular set of concerns derived from the topic at hand – cellular telephony – the desire to push visitors to understand their engagement and relation to the wider world whether configured as nature or culture, is core to the mission of NMNH.

To this end, through the collaborative dynamic of the exhibit design process, the team settled on the following core exhibition themes:

- Cellphones cannot be divorced from the natural world.
- Cellphones are seen everywhere, but the global network of people and infrastructure that supports them is often invisible.
- People use cellphones to connect and communicate with others in ways that are at once personal *and* cultural.
- The cellphone technology we create and use shapes us socially and shifts the ways we relate to each other.

Opening on 23 June 2023, the exhibition is situated in Hall 27 (5,700 square feet) on the museum's second floor between the natural history show, *Objects of Wonder*, and the *Bone Hall, Egyptian Mummies*, and the *Insect Zoo*. Hall 27 was chosen specifically because it offered a built-in educational studio that allows for exhibition-related programming.

The exhibition consists of the following sections (these titles are used internally by the museum team and are represented in the exhibit through a set of emoji titles): (1) 'Introduction,' (2) 'Resources,' (3) 'Infrastructure,'

(4) 'Cultural Uses,' (5) 'Circular Economy,' and (6) 'Influence.' I will turn to a discussion of each of these sections in turn.

Exhibition introduction

Understanding that the NMNH receives around four to five million visitors annually, and that the majority come to see the Hope Diamond and the renovated Dinosaur Hall during the one to two hours they typically spend in the museum, we wanted to create an arresting entry experience that would pull the visitors into the hall. Guided by the question "How do our cellphones connect us to the world?", we created what the team has come to refer to as the cellphone forest: four large-scale monitors fabricated to look like larger-than-life cellphones. The back of each monitor contains graphics to show the interior of a phone without the casing to help prompt visitors to understand that the exhibit is a metaphoric exploded diagram of a smartphone. To that end, the walls present a range of names from 20 languages that people use to refer to their device with the hope of bringing into view the global nature of the experience of cellular telephony. Each monitor runs through a series of displays that help preview contents of the show before shifting to a screen on which visitors see themselves and have their head replaced by a shifting array of emojis, depending on their facial expression. We hoped to create an experience that engages visitors to take images with their own smartphones. These monitors introduce visitors to the chatbot and data-visualization features of the show.

The group chat, which is a scripted web app that simulates a chatbot, guides visitors through a series of questions about their cellphone use and understanding of their device. Drawing on four fictional characters found in the *Influence* section, we created a set of conversations that were intended to provide visitors with points of further reflection and engagement. Cognizant that it would be odd for such an exhibition not to engage with the smartphones of visitors, this was a solution that allowed us to develop something both within budget and that would allow us to visualize people's responses later in the show. Throughout the exhibition there are monitors with which to interact so that people without a smartphone device can also participate. The group chat is also accessible through a QR code that directs them to the web app on their device.

Resources

Asking "What goes into your cellphone?", this section of the exhibition invites visitors to consider that everything in their device comes from the earth and is entangled with a particular place and set of activities and people (Arboleda, 2020; Smith, 2021). To help materialize this, a large display case designed to look like the top of a cellphone occupies the center of the space. Fifty-two minerals which give rise to 65 elements that are used to manufacture a typical cellphone are displayed on translucent plinths. Distinguishing between an element, a mineral, and a rock in the accompanying reader rail, the mineral and element specimens light up to indicate what portion of the phone – represented by Perspex cutouts of the cellphone's body, motherboard, battery, camera, and touchscreen – they are found in. I decided upon 65 elements to help visitors make the connection between the materials in their devices and the wider world. To help reinforce this point, we highlight how your body is composed of 24 elements from the periodic table's 118 known elements.[6]

Positioned around this central case are four other cases that display eight elements, each of which are examined from their form as a mineral to a component in a cellphone. These four cases each represent different sections of the cradle-to-cradle supply-chain of our devices: (1) extraction (tantalum and lithium), (2) design (aluminum and indium), (3) manufacturing (gold and silicon), and (4) reclamation (copper and neodymium). Each case identifies where that mineral can principally be found in the world and the processes used to turn it into a specific feature of the phone. Alongside each example is a full-scale portrait of an individual whose job is central to this process. Throughout the show these personal profiles were built out of an interview with the individual by the exhibit team or in some cases through a proxy. These people's quotes help to highlight a key aspect of their job, and their profiles put human faces onto the global supply chain of our devices (Bell et al., 2018b).

Infrastructure

Situated east of *Resources*, the *Infrastructure* section asks, "What does it take to make our cellphones actually work?" Here we present the ecosystem of terrestrial and marine network infrastructure through which our devices work. Displaying antennae, cell site, the portion of a 5G cellphone tower, a radio head, network router, and both undersea and terrestrial cables, we present together all of the apparatus that enables a call. A panel titled *Anatomy of a Call* walks through the steps and thus materials involved in making a call from Washington, DC, to Madagascar. We also explain spectrum and how it is a finite resource, and walk visitors through the power needed to run the data centers that are the backbone of the system. As in other areas of the exhibition, we have a 'future focus,' which examines the innovations that help to address a given problem in the industry.

Four personal profiles, including a Zapotec community organizer whose community built their own network (see González, 2020), and three engineers who work on 5G and spectrum respectively, help to bring into view different issues relating to connectivity, not only in terms of the science, but also the politics around access. Here we also touch upon how infrastructure builds upon older forms and how this is crucial to determining access (Durairajan et al., 2015). The section elaborates upon five generations of cellular technology and addresses the opportunities and anxieties around 5G (Mattern, 2019). As with other sensitive topics addressed in the exhibit – the conflict around minerals and health concerns relating to social media – the exhibition's goal is to present the science around a given topic, raise awareness of the issues at hand, and create a space for conversation.

Cultural uses

Across from *Infrastructure*, on the exhibit's western side, is the section devoted to *Cultural Uses*. Pushing against the notion that technology both flattens and is outside of culture, this section of the exhibition examines how cultural diversity is sustained, thrives and emerges anew through this technology. Here my curatorial intent was to set up comparisons that both make the familiar strange and the strange familiar, but to also open up conversations about cultural similarities and differences without exoticizing them. A central wall is divided into two large sections – 'Before Cellphones' and 'Because of Cellphones.' The former section funnels into a singular smartphone to convey the range of technologies (pay phone, concert tickets, VCR, TV, camera, books, etc.) that our devices have subsumed but have not necessarily replaced. The hope is that this array of 82 objects materializes for visitors the vast array of functions that have been collapsed into the cellphone and help convey their importance in people's lives. It also provides visitors with an opportunity to participate in discussions about the shifting role of technology and the ways that technology has helped shape our sense of self and the public domain.

The other section of the wall, 'Because of Cellphones,' is divided into five further sections – 'Connection,' 'Create,' 'Work It,' 'Say It,' and 'Save It,' in which different objects are clustered. It focuses on the ways in which the cellphone becomes a means of expressing one's identity through the containers that a person carries a cellphone in, including clothing, decorative cellphone cases (from pop art to Indigenous designs), and toy cellphones. Objects in the Connection section explore how we can connect to ancestors with examples of joss paper (Blake, 2011) meant to represent digital devices for one's ancestor; how technology can become a vehicle for the dead as seen with an abebu adekai ('receptacles of proverbs') in the form of a Nokia cellphone (Bonetti, 2009); and how cellphones can even extend our own senses with a digital microscope, a telephoto lens, and a stethoscope attachment. This section also demonstrates how people might want to evade connection through, for example, faraday pouch, and face paint intended to confuse facial recognition (de Vries & Schinkel, 2019). The Save It section features a display of a MPESA storefront with a soundscape of interactions in Nairobi, Kenya, in five different languages (English, Luhya, Kiswahili, Luo, and Kamba) (Kusimba, 2021). For this section we present personal profiles of a Kenyan businesswoman, and a Chinese student, both of whom use their device and mobile money. Elsewhere in this area, we feature the profiles of a Yolngu media maker living in Raymangirr, Northern Territory, Australia (Miyarrka Media, 2019), a student who successfully petitioned Unicode Consortium to create a hijab emoji (Ohlheiser, 2016), and two Myaamia speakers who use their devices to learn and sustain their Native American language (McCarty, 2018). These profiles help highlight diverse stories about the creative ways this technology is used. An adjacent theatre space allows visitors to see an edited selection of nine documentaries that examine cultural experiences along different aspects of the cradle to cradle experience of the cellphone around the world.

Circular economy

This section of the exhibition asks visitors to contemplate what the implications of their decisions once their device stops working or is no longer desired: putting it in a drawer, throwing it away, trading it, recycling it, or repairing and donating their device. Each of these topics is explored through a series of objects, personal profiles, and accompanying texts. The focus is on the trade-off of each decision and the implications of one's decisions as part of a wider ecology of action and inaction. Aware of both the misconceptions around e-waste and the idea that this is an issue that only the Global South deals with (Bell, 2019), we foreground an array of examples, from a copper burner working in Accra, Ghana, who sees his job as a way of advancing his community in Northern Ghana economically (Little, 2021), to employees at a recycling business in Rochester, New York. Drawing on collaborative ethnographic work carried out in Washington, DC, with third-party repair technicians (Bell et al., 2018b), the team developed a game about cellphone repair which is framed by a display of tools and the personal profile of a repair technician based in Washington, DC.

Our goal in this section was to promote awareness about the implications of one's actions and to help foster more care of and around the use of one's devices. Central to the space is a large sculpture of 465 cellphones that have been temporarily removed from recycling for display to help convey the scale of the e-waste issue on a local and global level. Along the three sides of the sculpture are animations that were developed with student interns at the Maryland Institute College of Art (MICA). These explore planned obsolescence, repair culture, and the future of sustainable cellphones. This sculpture has become a surprise star of the exhibition as visitors identify cellphones they once owned.

Influence

Shifting away from the display of objects, the *Influence* section features a large comic book mural designed in collaboration with comic book illustrator Khary Randolph and comic book writer Joanne Starer. My curatorial intention was to create a space in which a series of examples would prompt visitors to reflect on their own experiences. Informed by my collaborative ethnography carried out on the experiences of students of a Washington, DC, high school (Bell et al., 2021a; Dent et al., 2022) – as well as the work of others focused on issues of diversity, equity, and the politics of technology (boyd, 2014; Benjamin, 2019; Richardson, 2020) – the mural focuses on three teenage friends and their cat. The teens occupy different positions in their attitudes about technology, while the cat breaks the fourth wall and addresses visitors. The mural consists of three large panels that explore cellphone use in terms of the themes of connection, privacy, and health. In between these large panels are a series of displays that examine moral panic about different new media, the cellphone as witness, the cellphone as catalyst, cellphones and misinformation, bias in technology, and the cellphone as a distorting mirror.

It is here that visitors can interact with the data visualization that they are introduced to at the exhibition's entrance and which they are periodically reminded of throughout the show via QR codes and kiosks. This experience is intended to create further points of reflection about visitors' own experiences with their cellphones (all the data collected is anonymous). This interactive illuminates the changing realities of people's understanding and experience of cellular technology. In the effort to address agency in and with technology, we have a series of 15 portraits of people ('Changemakers') who are pushing mobile technology and social media to address issues of social and environmental justice and are working to reshape the technology. Through these individuals and their commentaries we hope to inspire visitors to think about and pursue STEAM careers.

Exhibiting unseen connections

In one of the exhibition's periodic reviews, a colleague commented that the show is experimental and encouraged the museum's executives present at the presentation to embrace this ethos and the promise of what the exhibition could do in terms of its interdisciplinarity and attempt to reach new audiences. He also cautioned that, as with all experiments, it might yield results that we could not anticipate. In the interceding years, I have come to appreciate and embrace my colleague's comments. As Anna Lowenhaupt Tsing reminds us, "big histories are always told through insistent, if humble, details" (2015, p. 111). The details focused on in this exhibit are the materials, people and places, and the relations between them that are materialized in our cellphones. *Cellphone: Unseen Connections* was intended to bring into view aspects of where our cellphones begin and end. While it remains to be seen what the exhibit will result in, one of my hopes for this experiment, it is that visitors will be inspired to think more about their relationships to their device (and technology more widely), to think about how they exist in mesh of relations that are not always acknowledged, between human and nonhuman others, and that they have agency in relation to technology.

Notes

1 Collectively our work has received grants from the George Washington University and Smithsonian Institution Opportunity Fund Research Program (2011), the Smithsonian Consortia for World Cultures and American Experience (2012, 2013), Wenner Gren (2012), National Museum of Natural History Web Advisory Grant (2013), Smithsonian Scholarly Studies (2013, 2017), and the National Science Foundation Senior Award (Grant #1534589, 2015).
2 Cynthia E. Smith of the Cooper-Hewitt, Smithsonian Design Museum, Joel Kuipers, and I convened this workshop on 28 February and 1 March 2013. The workshop was funded by the Smithsonian's American Experience and Valuing World Cultures Consortiums.
3 Joel Kuipers and I convened another workshop, *Linguistic and Material Intimacies of Mobile Phones,* which took place between 5 June–7 June 2013. Funded by the Wenner-Gren Foundation, in this workshop we explored the social and material implications of cellphones (see Bell et al., 2018a; Kuipers et al., 2018).
4 We have relied on the work of Pew Research Center in their Teens and Tech Series, as well as the work my colleagues and I carried out in Washington, DC, high schools to understand this demographic.
5 While drawing on scholarship relating to political ecology and materiality (Smith, 1990; Escobar, 1999; Tsing, 2015), I am mindful of how this exhibition builds, albeit with different intentions, on the long-standing anthropological tradition at the National Museum of focusing on technology, invention, and the environment (see Mason, 1894).
6 Here I was particularly influenced by a visit to the Royal Ontario Museum's Barrick Gold Corporation Gallery in 2015, which features a touch screen interactive that is designed for users to identify the minerals and elements in everyday objects. I visited the ROM as part of the Wenner-Gren funded workshop *The Anthropology of Precious Minerals* (Vallard et al., 2019).

Reference List

Anderhag, P., Wickman, P., Bergvist, K., Jakonson, B., Hamza, K. M., & Säsljö, R. (2016). Why do secondary school students lose their interest in science? Or does it never emerge? A possible and overlooked explanation. *Science Education, 100*(5), 791–813. https://doi.org/10.1002/sce.21231

Arboleda, M. (2020). *Planetary mine: Territories of extraction under late capitalism.* Verso Books.

Bell, J. A. (2017). A bundle of relations: Collections, collecting and communities. *Annual Review of Anthropology, 46*(1), 241–259. https://doi.org/10.1146/annurev-anthro-102313-030259

Bell, J. A. (2019). 'Check out that gold plated board!': The performance, poetics and dirty realities of scrapping cellphones and electronics in North America. In A. Vallard, A. Walsh, & E. Ferry (Eds.), *The anthropology of precious minerals* (pp. 43–68). University of Toronto Press.

Bell, J. A., & Geismar, H. (2009). Materializing Oceania: New Ethnographies of things in Melanesia and Polynesia. *The Australian Journal of Anthropology 20*(1), 3–27. https://doi.org/10.1111/j.1757-6547.2009.00001.x

Bell, J. A., Kobak, B., Kuipers, J., & Kemble, A. (2018a). Unseen connections: Introduction. *Anthropological Quarterly, 91*(2), 465–484. https://doi.org/10.1353/anq.2018.0023

Bell, J. A., Kuipers, J., & Dent, A. (2021). *How cellphones make and break human connections.* www.sapiens.org/culture/cellphone-ethnography/

Bell, J. A., Kuipers, J., Hazen, J., Kemble, A., & Kobak, B. (2018b). The materiality of cell phones repair: Re-making commodities in Washington, DC. *Anthropological Quarterly, 91*(2), 603–633. https://doi.org/10.1353/anq.2018.0028

Benjamin, R. (2019). *Race after technology: Abolitionist tools for the New Jim Code.* Polity Press.

Bennett, J. (2010). *Vibrant matter a political ecology of things.* Duke University Press.

Blake, C. F. (2011). *Burning money: The material spirit of the Chinese lifeworld.* University of Hawaii Press.

Bonetti, R. (2009). Absconding in plain sight: The Ghanaian receptacles of proverbs revisited. *RES: Anthropology and Aesthetics, 55*(1), 103–118. www.jstor.org/stable/25608838

boyd, d. (2014). *It's complicated: The social lives of networked teens.* Yale University Press.

de Vries, P., & Schinkel, W. (2019). Algorithmic anxiety: Masks and camouflage in artistic imaginaries of facial recognition algorithms. *Big Data & Society, 6*(1), 1–12.

Dent, A., Bell, J. A., & Kuipers, J. (2022). Cellular ambivalence in a digital age. *Anthropology Today, 38*(1), 18–20. https://doi.org/10.1111/1467-8322.12698

Durairajan, R., Barford, P., Sommers, J., & Willinger, W. (2015). Intertubes: A study of the US long-haul fiber-optic infrastructure. In *Proceedings of the 2015 ACM conference on special interest group on data communication* (pp. 65–578). https://doi.org/10.1145/2785956.2787499

Escobar, A. (1999). After nature: Steps to an antiessential political ecology. *Current Anthropology, 40*(1), 1–30. https://doi.org/10.1086/515799

González, R. J. (2020). *Connected: How a Mexican village built its own cell phone network.* University of California Press.

King Jr, M. L. (2010). *The trumpet of conscience.* Beacon Press.

Kuipers, J., & Bell, J. A. (2018). Linguistic and material intimacies of cell phones: An Introduction. In J. A. Bell & J. Kuipers (Eds.), *Linguistic and material intimacies of cell phones* (pp. 1–30). Routledge Press.

Kuipers, J., Bell, J. A., Kobak, B., Kemble, A., & Hazen, J. (2018). Intimate materialities in cell phone repair: Performance, anxiety and trust in DC repair shops. In J. A. Bell & J. Kuipers (Eds.), *Linguistic and material intimacies of cell phones* (pp. 237–265). Routledge Press.

Kusimba, S. (2021). *Reimagining money: Kenya in the digital finance revolution.* Stanford University Press.

Lei, R. F., Green, E. R., Leslie, S. J., & Rhodes, M. (2019). Children lose confidence in their potential to 'be scientists', but not in their capacity to 'do science'. *Developmental Science, 22*(6). https://doi.org/10.1111/desc.12837

Lepawsky, J. (2018). *Reassembling rubbish: Worlding electronic waste.* MIT Press.

Liboiron, M. (2021). *Pollution is colonialism.* Duke University Press.

Little, P. C. (2021). *Burning matters: Life, labor, and e-waste pyropolitics in Ghana.* Oxford University Press.

Mason, O. (1894). Technogeography or the relation of the earth to the industries of mankind. *American Anthropologist, 7*(2), 137–161. https://doi.org/10.1525/aa.1894.7.2.02a00000

Mattern, S. (2019). *Networked dream worlds: Is 5G solving real, pressing problems or merely creating new ones?* www.reallifemag.com/networked-dream-worlds/

McCarty, T. L. (2018). Community-based language planning: Perspectives from indigenous language revitalization. In L. Hinton, L. Huss, & G. Roche (Eds.), *The Routledge handbook of language revitalization* (pp. 22–35). Routledge.

Miyarrka Media. (2019). *Phone & spear: A yuta anthropology.* Goldsmiths Press.

Ohlheiser, A. (2016). *There is no hijab emoji: This 15-year-old student is trying to change that.* www.washingtonpost.com/news/the-intersect/wp/2016/09/13/there-is-no-hijab-emoji-this-15-year-old-student-is-trying-to-change-that/

Richardson, A. V. (2020). *Bearing witness while black: African Americans, smartphones, and the new protest# journalism.* Oxford University Press.

Smith, J. H. (2021). *The eyes of the world: Mining the digital age in the eastern DR Congo.* University of Chicago Press.

Smith, N. (1990). *Uneven development: nature, capital, and the production of space.* Blackwell.

TallBear, K. (2019). Caretaking relations, not American dreaming. *Kalfou: A Journal of Comparative and Relational Ethnic Studies, 6*(1), 24–41. https://doi.org/10.15367/kf.v6i1.228

Tsing, A. L. (2015). *The mushroom at the end of the world: On the possibility of life in capitalist ruins.* Princeton University Press.

Vallard, A., Walsh, A., & Ferry, E. (2019). *The anthropology of precious minerals.* University Toronto Press.

2 Lives on shelves

Constructing histories of computer in the museum store

Simona Casonato

The history of information processing naturally attracts the interest of large science and technology museums (STMs) with a broad remit. These museums typically keep collections of calculators, computers, networking hardware, and software.

The large part of the computing collections at the Museo Nazionale Scienza e Tecnologia Leonardo da Vinci (MUST) can only be viewed in the museum's storage – a short-term situation for some objects but a long-term condition for many others. In his seminal essay on the cultural biography of things, at one point Igor Kopytoff asked: "What of a Renoir ending up in a private and inaccessible collection? Of one lying neglected in a museum basement? [...] The cultural responses to such biographical details reveal a tangled mass of aesthetic, historical, and even political judgments, and of convictions and values that shape our attitudes to objects labelled 'art'" (Kopytoff, 1986, p. 67).

With rapid advancement in information technology, STMs continue to gather artefacts that are 'singularized,' to use again Kopytoff's terminology, as material evidence of the origins of our present digital age.[1] Given the range of hardware and software artefacts produced throughout the history of computing, the possibilities for exploration in the museum are almost endless (Alberti et al., 2020). Ordinarily, museums display only a small section of their collections. In the case of a vast phenomenon like the history of the 'digital,' therefore, there may be a significant discrepancy between exhibited and stored artefacts. Indeed, as the perception of technological relics of the past typically evokes images of obsolete and disposable items of technology, rather than a heritage to be preserved, the idea of a dusty basement may not sound so outrageous in this context (Durant, 2017, p. 38).

The concept of the museum is defined historically by a practice of collecting, but STMs have developed a peculiar attitude to this aspect of their mission. The function of STMs has mainly been discussed within the field of the public communication of science and technology, which has focused primarily upon exhibitions and educational or cultural programs, often overlooking the role of the collection per se (Spada, 2022). Notably, since the

DOI: 10.4324/9781003424703-4

1960s STMs have been judged as more successful at public engagement when they turn their attention away from 'idolized' historical objects to concentrate on the visitors' experience of contemporary science (Schiele, 2014). STMs are not perceived as purely historical museums: artefacts have always been seen as if they were a rhetorical device among other elements of exhibitions' storytelling; here the past seldom is the main topic (Bud, 2017). As a result, the meaning of science and technology collections has lost some focus, and curators are making attempts to reassess it (Alberti, 2017). This scenario should be taken into account when we reflect upon how the biographies of such artefacts are used to produce historical narratives. Samuel Alberti has argued that object biographies make the museum a meaningful "vessel for a bundle of relationships" (2005, p. 561), but how does this potential sit in the context of computing objects that lie dormant in the 'basement' of an STM? How are these objects able to participate in public history if they are not on display?

Until very recently, the cultural function of museum storage in society has attracted little theoretical reflection (Kreplak & Mairesse, 2021). As the twentieth-century progressed, the identity of museums has become increasingly defined by exhibitions rather than by collections (Poulot, 2018, p. 25). Non-exhibited objects have been perceived as discarded and "unloved" (Woodham et al., 2020). Popular narratives talk of dust and negligence, and the notion of the museum store is still associated with a range of negative perceptions as "the inverse of display," reminiscent of private and elitist cabinet of curiosities (Brusius & Singh, 2018). Recently, both perceptions and practices appear to be changing: stores have been rediscovered, but often as an alternative form of exhibition, while entire museums are (re)created in the fashion of a store room. Recent studies have explored the narrative potential of these 'paused' objects when they are cared for by different categories of museum users, such as the curator or "expert enthusiasts" (Geoghegan & Hess, 2015; Haines & Woodham, 2019).

Beyond their use for alternative forms of display, however, I would like to explore how objects in museum storage do not just sit but *live* as historical sources; they are active not despite but *because* of their condition as stored objects. To hone in on examples in a very focused way, I will consider two items in the MUST computing collections that are almost opposite in their 'rhetorical' features. The first artefact is an iconic machine from a remote past: an example of a Pascal's adder, or '*Pascaline*' (D776). The second artefact is a trivial piece of hardware from the recent past: a set of control panels or *plugboards* used in programming punched-card machines (IGB17478). Pascal's adder is a replica, while the panels are authentic. I confess I selected these two items quite randomly; however, once interrogated, the objects revealed some unanticipated historical 'kinships,' emerging from the symbolic condition of having been conceptualized, at some point in their existence, as belonging to museum collections.

Public and private faces of museum objects

The Pascaline is a seventeenth-century mechanical calculator that was invented in Rouen, France, by Blaise Pascal (1623–1662). The French philosopher and scientist was introduced by his father, Étienne, a finance intendant, to the complex task of calculating taxes. Beginning in 1642, Pascal spent ten years of his life conceiving one of the first machines capable of automated calculation.[2] Today, only eight original artefacts belonging to Pascal's project are known to have survived, but many replicas can be found in museums and private collections.[3]

The second artefact I selected from the museum's storage, the plugboards, are plastic trays with "a lot of holes, called hubs, into which you can insert wires with special tips" (IBM, 1956, p. 7) to program electromechanical punched-card machines. Such technology began to be developed at the end of the nineteenth century after Herman Hollerith's invention of a system to process data for the United States population census in 1890. During the 1910s, the development of Hollerith's method provided a starting point for the International Business Machines Corporation (IBM), which became very influential in the development of punched cards as a common data processing tool, right up until the last quarter of the twentieth century (Cortada, 2019).

In the previous two paragraphs I have provided the type of essential information that could be found on a panel in a typical exhibition about the history of computing. According to the traditional approaches adopted by STMs to historical narratives, objects shown to the public represent different steps in a historical progression of innovation. The Pascaline and the plugboards would both act as historically iconic components in a multimedia exhibition environment, but with different types of historical relevance. MUST has displayed a Pascaline on several occasions. In 1959, the temporary exhibition *Il calcolo automatico nella storia* (*Automated Calculation in History*) presented a traditional linear account of development, from the ancient abacus to technologies of the modern era. The booklet published to accompany this exhibition shows a different example of Pascaline, which was also a replica (IBM Italia, 1959, p. 11).[4] The more ambitious project of the permanent Sezione di Informatica [Informatics Gallery] created in 1975 saw the arrival of another Pascaline. After the dismantling of this exhibition in 2005, this Pascaline was transferred to a display case next to the entrance of the new hands-on laboratory on robotics. The visitors' activities designed for this space did not entail any special reference to the history of the artefact, which played a more 'decorative' than explicative role.[5] In 2012 the replica was moved to a temporary exhibition, *Tecnologie che contano* (*Technologies That Count*), where it assumed its original role of demonstrating a key breakthrough in computing history. In 2018 it was put into storage to make room for exhibitions featuring artefacts and themes that were understood to be more urgent.

The Pascal adder is considered a must-have artefact by SMTs, regardless of its authenticity. Like actors on the stage, replicas play the role of *the* Pascaline. At MUST the same historical plot was staged from 1959 to 2018, with some necessary updates and a growing sophistication in the use of exhibition techniques, but only minor variations in the basic conceptual structure. The substitution of the artefact between 1959 and 1975 suggests that it did not matter too much *which* actor performed the role.

Both exhibitions were sponsored and organized by IBM's Italian branch, which directly provided many objects for the exhibition: for the company, the 'fake' Pascaline proved to be effective in generating publicity, receiving attention in the press (Mondini, 1975). IBM also introduced to the 1975 exhibition a number of its own products. Several plugboards are given a brief description in a book issued to accompany the exhibition, providing brief notes on the discourse relating to punched-card machines, which bridged the era of electromechanical technology with a new age of much quicker electronic computers (De Pra', 1975, p. 27).[6] The plugboard set IGB17478 was donated more recently, but it has never been exhibited and has therefore not been prepared and interpreted for exhibition as yet.

Currently the Pascaline and the plugboards are kept in storage on identical shelves, which deprives objects of the visual and narrative hierarchies typical of the exhibition. In this context, we can encounter artefacts as more equal elements. Museum store rooms are like "a dictionary of words waiting to get placed in a sentence" (Shaw, 2018, p. 156). Here it is easier to see objects as they were "naked," like "facts without their clothes on," in all their raw, seductive material evidence (Marcus, 2012, p. 189). In the museum storage space, the effort to preserve artefacts is the central task. This has implications for the ways in which objects are encountered. On the one hand, the sight of collections *en masse* reinforces and gives substance to abstract notions of memory and history. The agglomeration of items emphasizes that the "historic weight" of museum collections is not just a metaphor (Warrior, 2018, p. 136). It also foregrounds the endeavor of collective memory creation and maintenance. On the other hand, the museum store also allows us to acknowledge that, in terms of object biographies, arrangements in storage and arrangements in exhibitions are entirely separate practices, but both provide interesting contexts or conditions for the artefact. In storage, objects are without their mask. Our Pascaline is neither an iconic adder nor just a replica: it is *that* replica. The plugboards are *those* plugboards.

The life of a copy

The manipulation of our Pascaline replica reveals the signature of its actual maker. Roberto Ambrogio Guatelli (1903–1993) was a Lombard artisan who led an adventurous life. After establishing an international career as a model maker and an expert on Leonardo da Vinci, in 1951 he was hired by the IBM president Thomas Watson Sr. and given the specific assignment of replicating ancient calculating artefacts. In making his Pascaline, Guatelli apparently mixed the features of two different original machines: the so-called "Louis Périer" (CNAM, Paris) and the "Marguerite Périer" (City of

Clermont-Ferrand).[7] IBM possessed an original in its private collections, but it is not entirely clear whether Guatelli copied from this example or reinterpreted it (Hénin & Casonato, 2020). Guatelli's Pascaline, although not necessarily unfaithful, seems accurate in its mechanical details: it is effective in reconstructing the "spirit" of early mechanical calculations and the experience of the seventeenth-century user (Agar, 1998).

If we focus on the artefact's status as a replica, and on its main commissioner, IBM, further historical insights emerge. The artefact participates in a cultural strategy deployed by the company that locates their origins and historical agency in the larger destiny of humanity, an attitude that has been customary in STMs (Canadelli et al., 2019) and has been adopted by contemporary big tech companies (Natale et al., 2019). The relationship with IBM has exerted an influence on the activities of MUST since its inception in 1953. Archival documents show the company as a customary partner, supporting the institution in many ways. IBM bought advertising spaces, loaned rooms to host corporate events, and supported exhibition projects through contributions of expertise and financial investment. Indeed, the IBM brand name was assigned to the hall hosting the 1975 exhibition. This relationship fostered the reproduction of past technologies to put on display: indeed, MUST possesses several artefacts made by craftsmen working for IBM.[8]

Looking closely at the biography of the Pascaline *as a replica*, then, we encounter some unexpected contemporary users of this artefact, such as IBM's communication officers. We also encounter less obvious domains of application, such as the dissemination in Italy of an IBM-oriented culture of computing, carried out via their cooperation with cultural institutions. In 1975 IBM and MUST chose *Informatics* as the title for the exhibition, rather than *Automated Calculation*. Created in 1962 and derived from the French term *informatique* (an elision of the words information and automatique), this term emphasized computers as a domain of data processing. Over the past decades in Italy, computers were often associated with scientific calculations in universities (Bonfanti, 2004). The new terminology is indicative of a process of normalization, projecting an everyday business perspective onto electronic computing.

The life of a component

The biography of the replicated seventeenth century calculator, here, shows an unexpected connection with the plugboards' story. When in 2019 a set of seven plugboards were offered to the museum, the accession was made on a taxonomic basis. The acquisition was meant to expand a collection of artefacts of the same age and made by the same producer, completing the set of punched-card machines acquired for the 1975 exhibition. On a purely technical level, a plugboard can be used to show how similar devices were adopted in the fields of telecommunication and data processing, or to contemporary eyes accustomed to digital black boxes – in other words, to machines that work without exhibiting their internal functioning, such as a contemporary laptop or smartphone – the plugboard shows a conflation between hardware and software, as material components storing instructions in the form of physical wired connections. The authenticity of the artefacts is a plus, since *any* plugboard can be an effective witness.

The encounter with plugboards in museum storage helps to expand narratives in different directions. Each item is marked with an adhesive label describing its computing task, such as the 'calculation of hours' or 'social deductions.' One is labelled 'IGE,' a tax applied in Italy between 1940 and 1972. Labels hint to specific users and periods, making the set an interesting testimony to the activities of Italian business administration over the past century. In June 2022, I met the individual who donated the plugboards, Mr. Luigi Marsi, a retired consultant from the field of business IT systems.[9] For Marsi, the plugboards evoked memories of his peculiar beginnings in programming in Italy when only a few specialists had heard about computers. In 1957, as a brilliant teenage student, he was selected to work in a new research center in Gallarate, a small town north of Milan, the *Centro per l'Automazione dell'Analisi Letteraria* (Centre for the Automation of Literary Analysis, CAAL), which was established in 1956 by the Jesuit priest Roberto Busa, one of the initiators of digital humanities.[10] Busa was undertaking monumental work in linguistics, indexing the opera omnia of Saint Thomas Aquinas (1225–1274) by using IBM punched-cards machines. In 1949 he travelled to New York and gained the direct support of the IBM president (Jones, 2018). Marsi recalled how he physically wired algorithms to produce the "lemmatized concordances" of medieval Latin,[11] gaining skills and knowledge that ensured him a fruitful career. By partnering himself with IBM, in the following years Marsi assisted many Italian firms in introducing punched-cars operations. The expertise he gained from Busa was transferred from one factory to another, in domains such as textiles and metallurgy. After 1968 Marsi switched the attention of his business to electronic computing. As for other former Busa operators, Marsi's experience suggests that IBM, by sustaining the priest's research, also indirectly produced technical expertise in automated calculation and information processing.[12]

IBM in northern Italy: elements of a cultural story

The biography of the replicated seventeenth-century calculator conceived by Pascal shows an unexpected connection with the story of plugboard technology. Both Pascal's adder and the plugboard have both featured as part of IBM's strategy to move beyond its own commercial activity and take part in cultural endeavors. Two artefacts that were randomly picked from storage at MUST show biographies full of connections to people and ideas going back and forth from distant places such as Manhattan and Lombardy in northern Italy, a region that experienced a process of fast industrialization in the second half of the twentieth century. The rise of IBM's influence in Europe was oriented idealistically, represented by the motto of its president: "World Peace Through World Trade." This commercial strategy was also to be given a vehicle through investments in collateral and apparently useless activities such as Busa's experiment in linguistics (Jones, 2019). The relationship of IBM with a national science museum produced benefits for both partners. MUST assisted IBM in spreading the good news of *informatics*, helping to root the commercial field of information processing in a historically and scientifically noble past. This action also seems to apply to the relationship with CAAL. Eventually, IBM's cultural politics underwent a kind of boomerang effect, donating a set of plugboards to MUST as a souvenir of the company's initiation into computing and their subsequent influence upon the world of technology.

Neither the Pascaline replica nor the set of plugboards would have existed today in the MUST collections without the direct and indirect cooperation of many actors: Watson Sr., Guatelli, Busa, Italian IBM employees, the donor Luigi Marsi, and of course the museum operators. Experiences with donors often confirms that technological objects "exert their holding power" over people because of the special circumstances in which these items entered their lives and marked momentous periods in their careers (Turkle, 2007, p. 8). Like scientific facts, museum objects are constructed as "solid meanings" (Volonté, 2003, p. 30). Over the years, all those different relationships and aims are stratified into physical items, marked by an inventory number.

Both the Pascaline and the plugboards potentially account for the diverse methods that Petrina Foti (2020) has examined in the work of curators facing the "unprecedented" case of the history of computer-based technology. With their mere presence, these artefacts allow us to *document* technological advancements in the field. The calculator can be *operated*, while the plugboards allow their user/donor to explain the practices of early programming. However, it is their unique identity that most *represents* a larger historical dynamic: the creation of the cultural conditions for the diffusion of automated information processing in everyday life and the naturalization of IBM's role in this landscape. Taking inspiration from these insights, a broader inquiry can be undertaken into other MUST collections that contribute to a social history of computerization in Italy.

This is made possible by the context of museum storage, which provides a place for artefacts' individual and 'local' circumstances to expand and become historically meaningful. According to anthropologist Thierry Bonnot (2009), the biography of a musealized object is special: the change of context enables the exploration of the multiplicity of all the subsequent contexts and human relationships in which it has been entangled, imbuing it with a variety of representations and practices that all inextricably belong to it and make it unique. Indeed, the Italian museologist Antonio Cirese (1977) dismisses the presumed 'death' of objects sitting in museum collections. The museum's task is to purposely separate things from the context of everyday life, to establish for them a different space and a different level of reality: musealized artefacts are documents *of themselves*. For stored collections at MUST, such considerations promote further steps in conceptualizing the role of objects in producing a public history of technology *beyond* their use in exhibitions. Their individuality as historical sources show their entanglement with society and culture and corroborate the understanding of intricate themes, such as the origins of the digital. As curators, we could learn the lesson of oral historians, who acknowledge the subjective and intentional nature of their sources – that is to say, people's accounts of their own lives – as historically meaningful. There is a specificity generated by the encounter between historians and their informants (Portelli, 1981). In our case, this is between museums and donors.

Lives are intentionally left on shelves. STMs do not need to be apologetic about the fact that things are in storage (Stevenson, 2018). Their cultural and social mission would not be complete if they did not offer the accumulation and care of collections, as a relevant resource with which to explore cultural phenomena in often unexpected ways: for example, what was labelled as a 'history of computing' in the past may generate new narratives in the future. After all, as Michel Foucault suggests, "the archive is first the law of what can be said" (1972, p. 129).

Acknowledgments

I am deeply grateful to Luigi Marsi, Luca Reduzzi, Laura Ronzon, Roberta Spada, and to the editors of this volume for their invaluable advice. I dedicate this paper to the memory of Silvio Hénin (1945–2022), who paved the way for a rigorous historiography of computer collections at MUST and whose contribution was fundamental.

Notes

1 I consider the history of computing in the larger context of the origins of 'digital,' a term that can be understood as defining the contemporary technologization and automation of communication and information. Terms in this domain are always replaced by new ones, as has happened to 'cybernetics' and 'information technology' (Kline, 2006).

2 Indeed, the mathematical precision sought by the Pascals played a significant role in the political history of taxation in France (Meurillon, 2018).

3 For instance, replicas can be found at the Science Museum in the UK, the Heinz Nixdorf Museum in Germany, the Ingenium in Canada, and both the Smithsonian Institution and the Computer History Museum in the United States. Originals are held in collections at the Conservatoire des Arts et Métiers (CNAM) in Paris, the Musée du Ranquet in Clermont-Ferrand, France, the SDK in Dresden, Germany, and in other private collections (Marguin, 1994, p. 61).

4 MUST attempted to obtain a Pascaline on loan from the CNAM in Paris but was presumably unsuccessful. The version put on display resembles other replicas that IBM sent to exhibitions around Europe, a reproduction of the Swedish queen Christina's Pascaline (Monnier, 2022). See Ucelli, G. (1958, June 23). *Letter to Vasco Ronchi*. Archivio Storico Museoscienza. Folder 1.10.4.8

5 Personal conversations with my colleagues, Summer 2022.

6 This was not a proper catalogue but a complementary visual essay on the history of computing: captions of artefacts and documents do not give any hint about their actual presence in the MUST gallery.

7 From the machine kept at the CNAM, Guatelli adopted the type of calculation (accounting), the handles and the smaller wheels on the upper part of the machine. From the other version of the device, he imitated the paper strip that reports units of measurement.

8 See Archivio Storico Museoscienza, Folders 1.7.16; 1.2.4.3.2.46; 1.10.4.8.

9 The following information comes from a personal conversation held on 30 June 2022.

10 Busa preferred to call his field "humanities computing."

11 As Steven Jones puts it, "all forms of a given word grouped under its dictionary entry or lemma" (2018, p. 1).

12 Other former CAAL employees reported similar work experiences, although female keypunch operators remained confined in this clerical role (Nyan & Passarotti, 2014).

Reference List

Agar, J. (1998). Digital patina: Texts, spirit and the first computer. *History and Technology*, *15*(1–2), 121–135. https://doi.org/10.1080/0734 1519808581943

Alberti, S. J. M. M. (2005). Objects and the museum. *Isis*, *96*(4), 559–571. https://doi.org/10.1086/498593

Alberti, S. J. M. M. (2017). Why collect science? *Journal of Conservation and Museum Studies*, *15*(1), 1–10. https://doi.org/10.5334/jcms.150

Alberti, S. J. M. M., Angus, S., Laurenson, S., Osborn, M., & Volkmer, L. (2020). *Digital collecting in museums: Approaches and opportunities, workshop report*. National Museums Scotland.

Bonfanti, C. (2004). Mezzo secolo di futuro: l'informatica italiana compie cinquant'anni. *Mondo Digitale*, *3*, 48–68.

Bonnot, T. (2009). L'approccio biografico alla cultura materiale. In A. Mattozzi & P. Volonté (Eds.), *Biografie di oggetti. Storie di cose* (pp. 27–36). Bruno Mondadori.

Brusius, M., & Singh, K. (2018). *Museum storage and meaning: Tales from the crypt*. Routledge.

Bud, R. (2017). Museums theme – adventures in museology: Category building over a century, and the context for experiments in reinvigorating the science museum at the turn of the twenty-first century. *Science Museum Group Journal*, *8*(8). https://doi.org/10.15180/170809/001

Canadelli, E., Beretta, M., & Ronzon, L. (2019). *Behind the exhibit: Displaying science and technology at world's fairs and museums in the twentieth century*. Smithsonian Institution Scholarly Press.

Cirese, A. M. (1977). *Oggetti, segni, musei sulle tradizioni contadine*. Einaudi.

Cortada, J. W. (2019). *IBM: The rise and fall and reinvention of a global icon*. MIT Press.

De Pra', R. (1975). *Tre secoli di elaborazione dei dati*. IBM Italia.

Durant, J. (2017). A Tape Measure and a "T" Stop. Or, Why Museums of Science and Technology Should Collect More Contemporary Artifacts. In A. Boyle, J. Hagmann *Challenging collections: approaches to the heritage of recent science and technology* (pp. 24–39). Smithsonian Institution Scholarly Press.

Foti, P. (2020). *Collecting and exhibiting computer-based technology: Expert curation at the museums of the Smithsonian Institution*. Routledge.

Foucault, M. (1972). *The archaeology of knowledge and the discourse on language*. Pantheon Books.

Geoghegan, H., & Hess, A. (2015). Object-love at the science museum: Cultural geographies of museum storerooms. *Cultural Geographies*, *22*(3), 445–465. https://doi.org/10.1177/1474474014539247

Haines, E., & Woodham, A. (2019). Mobilising the energy in store: Stored collections, enthusiast experts and the ecology of heritage, *Science Museum Group Journal*, *12*(12). https://doi.org/10.15180/191207/001

Hénin, S., & Casonato, S. (2020). Fake but true: Model maker Roberto Guatelli, science museums and replicated artifacts of computing history. *IEEE Annals of the History of Computing*, *42*(2), 20–32. https://doi.org/10.1109/MAHC.2020.2990452

IBM. (1956). *Functional wiring principles: IBM punched card data processing equipment*. International Business Machine Corporation.

IBM Italia. (1959). *Il calcolo automatico nella storia*. Tipografia Grisetti.

Jones, S. (2018). *Roberto Busa, S. J., and the emergence of humanities computing: The priest and the punched cards*. Routledge.

Jones, S. (2019). Foreword. In M. Passarotti & J. Nyhan (Eds.), *One origin of digital humanities: Fr Roberto Busa in his own words* (pp. xiii–xvi). Springer.

Kline, R. (2006). Cybernetics, management science, and technology policy: The emergence of 'information technology' as a keyword, 1948–1985. *Technology and Culture*, *47*(3), 513–535. 10.1353/tech.2006.0184

Kopytoff, I. (1986). The cultural biography of things: Commodification as a process. In A. Appadurai (Ed.), *The social life of things: Commodification in a cultural perspective* (pp. 64–92). Cambridge University Press.

Kreplak, Y., & Mairesse, F. (2021). The submerged part of the iceberg. *Museum International*, *73*(1–2), iv.

Marcus, J. (2012). Towards an erotics of the museum. In S. Dudley (Ed.), *Museum objects: Experiencing the properties of things* (pp. 188–201). Routledge.

Marguin, J. (1994). *Histoire des Instruments et machines à calculer*. Hermann.

Meurillon, C. (2018). Le chancelier, les nu-pieds et la machine: Pascal père et fils à Rouen. In J. Cléro (Ed.), *Les Pascal à Rouen, 1640–1648* (pp. 89–105). Presses universitaires de Rouen et du Havre.

Mondini, A. (1975, April 4). Dalla Pascaline al super computer. *Il Mattino.*

Monnier, V. (2022). Pascal's calculators: Distinguishing originals from replicas. http://ami19.org/Pascaline/IndexPascaline-English.html

Natale, S., Bory, P., & Balbi, G. (2019). The rise of corporational determinism: Digital media corporations and narratives of media change. *Critical Studies in Media Communication, 36*(4), 323–338. https://doi.org/10.1080/15295036.2019.1632469

Nyan, J., & Passarotti, M. 2014. (2022). *Interviews with Giuseppina Brogioli and Bruna Vaneili.* Reconstructing the First Humanities Computing Center. www.recaal.org/pages/interviews.html

Portelli, A. (1981). On the peculiarities of oral history. *History Workshop Journal, 12*(1), 96–107. https://doi.org/10.1093/hwj/12.1/96

Poulot, D. (2018). *Musei e museologia.* Jaca Book.

Schiele, B. (2014). Science museums and centres: Evolution and contemporary trends. In M. Bucchi & B. Trench (Eds.), *Routledge handbook of public communication of science and technology* (pp. 40–57). Routledge.

Shaw, W. (2018). Preserving preservation: Maintaining meaning in museum storage. In M. Brusius & K. Singh (Eds.), *Museum storage and meaning: Tales from the crypt* (pp. 152–168). Routledge.

Spada, R. (2022). Science and technology museums meet STS: Going beyond the galleries and into the practices. *Tecnoscienza: Italian Journal of Science & Technology Studies, 13*(1), 129–146. www.tecnoscienza.net/index.php/tsj/article/view/497/306

Stevenson, A. (2018). Lying in wait. Inertia and latency in the collection. In M. Brusius & K. Singh (Eds.), *Museum storage and meaning: Tales from the crypt* (pp. 231–239). Routledge.

Turkle, S. (Ed.) (2007). *Evocative Objects. Things We Think With.* The MIT Press.

Volonté, P. (2003). *La fabbrica dei significati solidi. Indagine sulla cultura della scienza.* Franco Angeli.

Warrior, C. (2018). Home from home: Memory and history, families and museums. In M. Brusius & K. Singh (Eds.), *Museum storage and meaning: Tales from the crypt* (pp. 131–139). Routledge.

Woodham, A., Hess, A., & Smith, R. (2020). *Exploring emotion, care, and enthusiasm in 'unloved' museum collections.* Amsterdam University Press.

Provocation no. 1: imparting the history of 'intangible things'

Mai Sugimoto

The history of science and technology is often explored as a means of education, a tradition that began in the nineteenth and early twentieth centuries and continues today. As a result, the history of science and technology including the history of computing is, in part, inherently for nonexperts. The outcomes of historical research are not solely intended for historians. They are often repurposed as teaching materials for students, reading matter for general enjoyment, or as museum exhibitions for the general public. However, imparting the history of 'intangible things' to those without associated specialized knowledge can pose significant challenges.

Nonexperts often lack knowledge of the relationship between scientific expertise, technological artefacts, and society. The difficulties of describing the various aspects of the history of science and technology in materials simplified for educational purposes are inevitable, and the question of how to address these difficulties has been widely debated.

The history of computing constantly grapples with these same issues. While it began with a focus on hardware developed in the US and UK, the academic field has since expanded to encompass global issues of maintenance, infrastructure, software, and social aspects such as policy, culture, and gender. In contrast, the emphasis on 'world's first' inventions and stories of 'geniuses' as inventors and entrepreneurs has often overshadowed the complexity of the field for the general public.

These problems can be particularly pronounced when dealing with the history of 'intangible things' in museums. For example, when describing the theoretical history of computers, such as the history of computer science, there may be few physical artefacts to display. In such cases, the individuals who devised the theories or the 'geniuses' who represent the field may be given disproportionate emphasis, resulting in the neglect of non-central actors and the underestimation of the dynamics of a broader group of stakeholders.

One example of this is the history of artificial intelligence (AI), which for years contained many theoretical aspects that were not put into practical use. The early history of AI is primarily a theoretical history, and when describing it for the general public, only a limited number of representative

DOI: 10.4324/9781003424703-5

figures are typically introduced, such as well-known mathematicians like Alan Turing, or AI researchers such as John McCarthy and Marvin Minsky who attended the Dartmouth Meeting in 1956. Due to the lack of users in early AI, it is challenging to give coverage to the issue of gender and race as it relates to the history of computing. For instance, when I oversaw the series of cover illustrations for the journal of the *Japanese Society for Artificial Intelligence* in 2023, which depict the history of AI, my collaborative supervisors and I encountered this problem. In this case, the illustrator attempted to alleviate the difficulties by incorporating female programmers in the depiction of the ENIAC project, the precursor of AI. Such difficulties can arise not only in the history of computing but also in the portrayal and exhibition of technological heritage in general.

For scholars of the history of computing, countering the prevailing historical perspective of 'exceptional geniuses' can often be arduous. These issues may not be resolved solely within a single exhibition or item of teaching material. Instead, a more comprehensive approach, which considers the history of museums and current museum practice, as well as the history of science and technology education, could provide an understanding of the various aspects and complexities of scientific and technological developments, raise awareness of these problems, and hopefully lead to potential solutions.

Part II

The life inscribed on computer technology

3 Restorations, replicas, and emulations in a museum of computing

Martin Campbell-Kelly and Mark Priestley

The display of historic computers in museums presents a number of curatorial challenges. Like many other technical artefacts, computers are functionally complex machines that are operated through intricate and often idiosyncratic interfaces. It is therefore difficult for a static display to communicate an understanding of how they worked or were used. To address this, museums of science and technology often display reconstructions or models of historic artefacts in operation. The expectation is that seeing such devices at work will enable visitors to better understand their design, behavior, and use.

Two things complicate the application of this practice to the case of computers. Firstly, direct and ongoing human interaction with computers, either as programmers or users, is crucial to their use. To fully appreciate historic machines, that is to say, not only the machine but the experience of using it need to be preserved (Keramidas, 2015). Secondly, much of the functionality of computers is invisible, as a result of the use of electronics and miniaturization, with the result that even a working exhibit reveals little of what is actually going on.[1] While the operation of historic computers can undoubtedly offer an 'aura' of authenticity, complementary approaches are needed to more fully explain the machines to visitors.

In this chapter, we examine how a range of approaches are applied in the UK's National Museum of Computing (TNMoC). The acquisition, preservation, and demonstration of historic computers in working order is central to TNMoC's mission to "bring to life the history and ongoing development of computing" (National Museum of Computing, 2022a, para.1), and we describe some central examples that illustrate the range of practices and activities that this involves. We identify three main approaches to the task of bringing historic machines to life. The first, *restoration*, applies when a machine has survived in a condition that is good enough to make it feasible to restore it to working order. This generally requires a process of repair and on-going maintenance that can be extensive, depending on the condition of the surviving artefact. Many significant historic machines have not survived, of course, but it may be the case that enough information has survived about a machine's design and construction to make it possible to create a physical replica of it, a

DOI: 10.4324/9781003424703-7

process that we call *rebuilding*. Both restoration and rebuilding are carried out with the aim of preserving material fidelity to the original, by using original components where they are still available, for example. A crucial difference between restoration and rebuilding is that, as in traditional museum displays, the former commands the authenticity and mystique of an original artefact (Agar, 1998).[2] It is often impractical or uneconomic to physically rebuild a machine, but the abstract nature of computation means that an *emulation* can still convey a good deal about how it worked. In the majority of cases, a computer's users are unaware of how the hardware functions and interact with the machine through a constrained user interface. An emulation is a replication of this user interface in software and can deliver an illuminating – and in many cases very convincing – experience of what it was like to use an old computer.

In this chapter, we describe significant examples of these three categories and explore how their presentation both in TNMoC and online contributes to and enhances the display and understanding of digital technology. In conclusion, we draw on the lessons learned at TNMoC to address the following question: what do restorations, replicas, and emulations add to the museum experience?

Restorations and working machines

TNMoC's unique selling point is the display of working computers. Some of these arrived needing extensive restoration after many years in storage, while others arrived having been recently decommissioned and in need of only a gentle coaxing back to life. One of our most extensive restorations, and one of our most important exhibits, is the Harwell Dekatron Computer.

The Harwell Dekatron Computer is believed to be the oldest working digital computer in the world, and it has had an eventful life. It was originally developed for the UK's Atomic Energy Research Establishment at Harwell, where it went into service in 1952. It was used for one-off jobs, including problems in the design of the first atomic reactor. However, by 1957 it was rarely used, having been largely replaced by a more modern machine, and was transferred to the Wolverhampton and Staffordshire College of Technology, where it was used to support the teaching of programming. In 1973, it was moved to the Birmingham Museum of Science and Industry, where it was displayed (although not operated) until the museum closed in 1997. It was then placed in storage. However, in 2008 it was recognized in a photograph by a TNMoC trustee who had visited the museum and had admired the machine 30 years earlier. The computer was then acquired by TNMoC, where it was restored and put back into operation in 2012 (Murrell & Holroyd, 2013). Since then it has been a centerpiece of the museum's First Generation Gallery, where it is displayed alongside some of its contemporaries, such as the Hollerith Electronic Computer (HEC), and a rebuild of the Electronic Delay Storage Automatic Calculator (EDSAC).

The Harwell Dekatron Computer's long and eventful biography presents a useful complement to familiar stories of invention and innovation. Built for reliability rather than speed, its design was conservative even for 1950, and its second career, as a teaching machine, was considerably longer than its first. Fortuitously, it is well suited for pedagogical purposes: it has the same basic design as all other computers but its construction exposes its basic components in a way that makes the structure of that design very accessible to viewers. Individual digits are visible as illuminated positions on the Dekatron tubes, and the slow speed of the electromechanical relays allows visitors to experience in a very direct and physical way the basic steps in computation that are invisible in more advanced all-electronic designs such as the EDSAC. The machine is still a valuable teaching tool: it is regularly in operation at TNMoC and used to demonstrate working programs and the principles of programming and program execution to visiting groups of school and college students.

The Large Systems Gallery houses several working mainframe computers. For example, we have a 1950s Marconi TAC computer that once controlled a nuclear reactor. This forms the basis of a fascinating display that explains the real-world application of an early computer. We have a second-generation Elliott 803 computer, a 1960s transistorized core-memory machine that was once used in a secondary school. This educational experience is being recreated for a new generation so that visitors can run their own Algol programs. Moreover, we have a large ICL 2966 mainframe dating from the mid-1980s, one of the most powerful of the ICL 2900 series. The processor sits alongside a series of disc stores – each the size of an automatic washing machine – helping to vividly convey the scale of an early computer. In another gallery the museum has put on display a wide range of desktop computers and games consoles from the 1970s and 1980s. As well as standard IBM-compatible PCs and Apple Macintosh computers, the gallery contains a full range of British machines from the microcomputer boom of the 1980s. Manufacturers include Sinclair, Amstrad, Acorn, and several less-known, no longer existing producers.

Many of these machines constitute fully working, 'hands-on' exhibits for visitors. It is fascinating to see middle-aged visitors reliving the computing experience of their youth and attempting to explain the systems to their children. In another exhibition space within the museum are located 18 BBC microcomputers set up in a staged 1980s-style classroom. This has proven a major attraction for school visits, enabling today's students to experience what computer education was like during the period in which their parents were raised. Indeed, TNMoC hosts school visits approximately 250 days during the year.

For all our working exhibits, ongoing maintenance is managed by volunteers. Our older mainframes typically arrived with spare circuit boards and maintenance documentation, and it is possible to replace discrete components on printed circuit boards. We are aware that for mainframes of the 1990s and beyond, which contain custom chips, this will be much less straightforward. That is a challenge for tomorrow's volunteers, and we have every confidence they will find a way. In the case of microcomputers, we aim to keep a minimum of three machines: a machine in the permanent collection for reference, a machine on display for hands-on use, and a third machine for the supply of spare parts. In the case of the BBC microcomputers, we keep a very large stock of these machines – enough for the foreseeable future; indeed, so many of these robust machines were manufactured that we have had to politely refuse further donations.

Rebuilds and replicas

TMNoC has on display two rebuilt historic electronic digital computing machines dating from the 1940s: the Colossus and the EDSAC. We prefer to use the term *rebuild* rather than *replica* because neither the original machines nor detailed plans or circuit diagrams exist for these machines. Consequently, their rebuilding has involved working with original plans and images that are fragmentary, and the use of creative engineering to design parts of the machines for which no specifications exist. By contrast, we use the term replica for the reproduction of a machine that still exists or for which there are detailed plans and pictorial evidence.

The historical 'Colossi' were a series of ten machines built between 1943 and 1945 by the UK's General Post Office to assist the Bletchley Park code-breakers' attack on the German Lorenz cipher. These machines represent the first large-scale application of electronic technology to computing. The first Colossus was completed by January 1944. Beginning in July 1944, a series of significantly more capable 'Mark 2' machines were constructed at roughly monthly intervals until the end of the war. All except two of the machines were subsequently destroyed, and knowledge of the Colossi was extremely restricted. Their existence only became widely known in the mid-1970s (Gannon, 2006; Copeland, 2006).

The Colossus rebuild project was led by TNMoC's founder Tony Sale (Sale, 1998, 2005). In 1989, Sale became secretary of the Computer Conservation Society (CCS), a specialist group of the British Computer Society set up to provide a focus for the conservation and restoration of historic computers. By 1991, Sale was leading efforts to save the historic Bletchley Park site from developers and had become convinced that it would be possible to rebuild Colossus using original components. Starting with only a handful of historic photographs and fragmentary circuit diagrams, Sale made use of contacts at GCHQ he had made while working for MI5 in the early 1960s, as well as information obtained from surviving GPO engineers who had worked on the original machines. He filled in gaps in the original plans by referring to details of other machines built by the GPO for Bletchley Park, reasoning that under wartime pressure the engineers would have reused circuits to perform similar functions on different machines.

By 1996 the rebuild team had achieved a working 'basic Colossus.' In preparation for its public unveiling, the team switched their attention to creating a closer resemblance to period photographs of the machine. The project was then energized by a substantial declassification in 1995 of wartime documents by the United States' National Security Agency. This included significant material relating to the decryption of the Lorenz cipher and was followed by the release of important new material by GCHQ. As a result, Sale decided that it was possible to attempt a more complete rebuild of a Mark 2 Colossus. This was achieved by 2004, when the rebuild was used to demonstrate one of the key cryptanalytical tasks that the original Colossi would have performed.

Each Mark 2 Colossi was slightly different in build, customized in some cases for specialized cryptanalytical tasks. Sale chose to build a machine that was as far as possible faithful to the appearance, design, and componentry of the wartime Colossi and which could carry out the central functions of those machines. However, the rebuild is not an exact replica of any one historic machine. It is perhaps best understood as simply the latest in the series of Colossi. Installed at TNMoC in an original wartime 'Colossus Room,' the machine is in regular operation. Thanks to Sale's attention to detail, the Colossus rebuild gives an extremely convincing impression of the appearance of the wartime originals.

The world's first practical electronic stored-program computer, the EDSAC, was first demonstrated at the University of Cambridge on 6th May 1949.[3] The EDSAC Replica Project was established as a registered charity in 2012 (Herbert et al., 2017)[4] and was funded by computing entrepreneurs and businesspeople in the Cambridge locale, often referred to as 'Silicon Fen.' The project trustees made an agreement with TNMoC to house the machine while it was under construction and to display it indefinitely when completed. The construction team initially consisted of about 20, mostly retired, electronic engineers who had spent their careers during the era of the transistor. This required them to learn vacuum tube electronics from textbooks and from the older members of the team with experience of the vacuum tube. The original EDSAC was constructed using vacuum tubes widely deployed in military equipment during World War II and which remained in use well into the 1960s. These are still available from electronics dealers, and enough were obtained for the 3,000 used in the EDSAC, including spares for potential use in the future. However, it is unlikely that there exist sufficient stocks on the open market to build another EDSAC. The original EDSAC used mercury-delay-line memory technology. Due to safety risks and sheer costs, mercury could not be used in today's museum environment. Instead, the technology of nickel-wire delay-line memory from the 1950s was used. This was in keeping with the spirit of the era and compatible with the EDSAC's serial mode of operation. Passive components such as resistors and capacitors manufactured in the 1940s no longer exist in any quantity, so newly manufactured components, electrically indistinguishable from those of the postwar era, have been used. The project began with fragmentary plans: there were just a few circuit diagrams and overall block diagrams. However, three years into the project a set of plans of the machine as it existed in 1951 came to the surface. These had been rescued from disposal by a technician at Cambridge University and kept safe for more than 50 years. This made the project of continuing construction more straightforward and also showed that those parts of the reconstruction that had been improvised were well-judged selections.

The EDSAC rebuild is taking far longer to complete than the original EDSAC. We believe this is because we have relied on part-time engineers (typically working just one day a week) with the inevitable continuity and communications problems. However, the construction has been a popular attraction in the museum. Visitors are fascinated to see the work in progress and to interact with the engineers. At the time of writing completion appears to be approximately six months away, although that may still be true in six months' time.

Simulations and emulations

In the museum context, a simulator or emulator can be seen as a 'virtual rebuild.' A simulator replicates the *logical* behavior of a computer but not its physical form. Simulators can take several forms, as software programs, web-based applications, or hardware-based systems using field-programmable gate arrays (FPGAs).[5] Simulators exist for most of the working machines at TNMoC. Here we focus on those for the Colossus and the EDSAC.

The first phase of the Colossus rebuild demonstrated basic functionality, but questions remained about how a number of other circuit boards connected together. Tony Sale resolved these questions not by further investigation of the hardware but by examining how the machine was used. He wrote a simulation called Virtual Colossus to reproduce the code-breaking procedures revealed in newly declassified sources. This provided crucial insights into the physical design of the machine.[6] Virtual Colossus consisted of an interactive simulation of the machine that was written in the then-current version of Javascript. Although drawing on knowledge and experience gained in the rebuild project, it is a separate artefact from the physical rebuild not only materially but also in functionality. Sale (2004) noted in particular that Virtual Colossus omitted the so-called 'spanning and rectangling gadget' present on the physical machines. At the same time, the Virtual Colossus took advantage of its software platform to add functionality for the benefit of its users, including a 'Test and Trace' switch that had no equivalent on the original Colossi.

Sale's Virtual Colossus can still be made to run, but its user interface now appears rather dated. In an interesting twist, the simulation itself has been rebuilt by computer programmer Martin Gillow (2022a). This "rebuild of the virtual rebuild," in Gillow's words (2022b), preserves Sale's back-end code but updates the user interface to offer a three-dimensional simulation of Colossus. Like Sale, however, Gillow has added convenient features such as a 'speed switch' that allows the modern user to run the simulation at the same speed as the historic machine, or to take advantage of the speed of modern computers. Like Sale's switch, this is presented as if it were a physical switch on the simulated machine, rather than a parameter or control of the framing simulation.

In the case of Colossus, then, the relationship between physical and virtual artefacts is far from straightforward. Sale's original Virtual Colossus was used as an exploratory design tool and was known not to be a completely faithful simulation of the historic machines. The more recent simulation combines a faithful recreation of Sale's code with a visual simulation of the physical Colossus rebuild. The value of these simulations is twofold. Firstly, they provide an animated illustration of a physical machine and so allow users to gain some sense of how the machines operated and perhaps even the experience of using them. This is of course heavily mediated by the characteristics of the simulators' user interfaces but is perhaps better than nothing. Secondly, as functional reconstructions of Colossus, the simulations support explorations and recreations of the code-breaking practices that were developed for the machines. Sale provided a suite of examples with his original Virtual Colossus, setting puzzles for his users based on original encrypted material.

Simulators have been used both in the development of the EDSAC rebuild and will also be used in future programming workshops. Throughout the EDSAC's construction, the logical and electronic functioning of the machine have been tracked using a FPGA-based simulator. The simulator mimics the behavior of the EDSAC at the component level, mirroring every single electronic signal in the system. The simulator enables circuits within the system to be verified before actual assembly and helps the resolution of complex timing conflicts. This has been essential in filling the gaps of the design information passed down to us. Without simulation this could only be achieved through a lengthy trial-and-error process. Because this emulation has been created on such a basic level, it can only be practically achieved by using a specialized FPGA system.

Besides being the first practical stored-program computer, the EDSAC was blessed with a particularly elegant and simple programming system that has enabled numerous simulators to be developed around the world (Campbell-Kelly, 2000). These have been used both by hobbyists and in computer science education, with both software programs and web-based simulators existing for each. Because modern computers are at least a million times faster than the original EDSAC, software-based simulators can run programs much faster than the original. Programs can be written, debugged, and run in a couple of hours, compared to the several days it could take using the original machine. For TNMoC, an EDSAC simulator has been developed that runs on popular computer platforms. When the rebuild is completed, the intention is to hold programming workshops in the spirit of the summer schools run by Cambridge University in the 1950s. The simulator will enable participants to learn to program the EDSAC on their personal computer prior to the actual workshop, where they will then be able to run their programs 'for real.' The original Cambridge University summer schools, it should be noted, were a fortnight in duration. Using the simulator it will be possible for participants to get the full experience in a workshop lasting only a day or a weekend. Despite the telescoping of the experience from days into hours, we believe our work-shops will constitute an authentic experience for those wanting to appreciate what it was like to be a programmer in the early days.

We know that our rebuilds and restorations cannot be kept running for ever. Twenty or fifty years into the future they may be beyond repair due to physical deterioration or the lack of available components to carry out such repairs. It is possible that our successors will decide to place a simulator within the exhibit that will enable the illusion of a functioning machine to continue. The museum would of course always acknowledge this artifice, but we think our visitors would always prefer to see a machine 'in steam' rather than as a static exhibit.

Conclusions: material, social, and discursive dynamics

The previous three sections have illustrated the variety and range of working exhibits of historic computers at TNMoC. The display of original artefacts alongside rebuilds and reconstructions is common to many science and tech-nology museums, but we have found the use of simulators and emulations useful to address the issues of interaction and 'invisible functionality' that are characteristic of computers. Both within and beyond the museum's walls, the exhibits enrich visitors' experiences and understanding of the practices of computing. For example, the original Harwell computer and the EDSAC rebuild share the same exhibition space, highlighting the difference between original artefacts and rebuilds and raising questions about the preservation of knowledge and technological change across generations. Rebuilds also allow a greater range of machines to be exhibited, placing a range of contemporane-ous machines together to give visitors a sense of the ecosystem of comput-ing at particular points in time. Although visitor engagement with historic machines at TNMoC is often limited to observation, the existence of simula-tors and emulations enables visitors to deepen their experience and apprecia-tion of the machines.

The activities described in this chapter have delivered two significant benefits to TNMoC. The construction, maintenance, and operation of the various classes of artefact have been an enormous stimulus to museum staff and volunteers, and the ability to experience historical computers in action greatly enhances visitors' experience. Historical computers never arrive at TNMoC with complete documentation. The processes of restoration and rebuilding therefore demand significant historical research and often elicit and generate lost knowledge about the design and construction of the machines, or the craft and technical skills necessary to employ that knowledge. The construction of an accurate simulation typically makes similar demands, at least on a functional level. Furthermore, in some cases writing a simulation can serve as an analytic tool to explore aspects of a machine's behavior and help fill in the gaps of documentation.

TNMoC is careful to distinguish the original machines on display from the rebuilds, but typically both classes of machine are displayed in similar ways. The regular display of these working machines requires volunteers to learn how to keep them in working order and to operate them. TNMoC's mission to 'bring history to life' therefore not only preserves the physical artefacts in good working condition but also replicates the historic practices of maintenance and use and the associated technical skills on which the machines depended in their original contexts. TNMoC was founded in 2005, so it is now a relatively mature institution. With this maturity has come the realization that our restorations and rebuilds will outlive their original engineering teams. We are now establishing a two-part program to ensure we can continue to maintain our major artefacts as working exhibits. Firstly, we are capturing and formally documenting the tacit knowledge of the original engineering teams in the form of engineering and maintenance manuals. Secondly, we are establishing an apprenticeship scheme so that new volunteers can be inducted into the ongoing maintenance and demonstration of our working machines.

The existence of working machines highlights a further issue, namely, the importance of the preservation of historic software to run on them. It is beyond the scope of this chapter to explore the complex issue of software preservation in any depth, but we note that the existence of working machines – physical or virtual – provides a useful testbed for the examination and preservation of historic software.

Notes

1 David Link has highlighted this aspect of what he calls 'algorithmic artefacts' (2016).
2 For a comprehensive discussion on the value of working artefacts, see also National Museum of Computing (2022a).

3 An outline history of the EDSAC features in virtually every book on the history of the computer. See, for example, Chapter 1 in Haigh and Ceruzzi (2021).
4 The project website can be found at National Museum of Computing (2022b).
5 Examples of several practical simulators for historic computers appear in Rojas and Hashagen (2000).
6 For a description of the simulation, see Sale (2004).

Reference List

Agar, J. (1998). Digital patina: Texts, spirit, and the first computer. *History and Technology*, *15*(1–2), 121–135. https://doi.org/10.1080/07341519808581943

Campbell-Kelly, M. (2000). Past into present: The EDSAC simulator. In R. Rojas & U. Hashagen (Eds.), *The first computers: History and architectures* (pp. 397–416). MIT Press.

Copeland, B. J. (2006). *Colossus: The secrets of Bletchley Park's codebreaking computers*. University Press.

Gannon, P. (2006). *Colossus: Bletchley Park's greatest secret*. Atlantic Books.

Gillow, M. (2022a). *Virtual Colossus: Bringing the world's first electronic digital computer into the 21st Century*. www.virtualcolossus.co.uk/

Gillow, M. (2022b). *The rebuilding of Colossus*. www.virtualcolossus.co.uk/rebuild.html

Haigh, T., & Ceruzzi, P. E. (2021). *A new history of modern computing*. MIT Press.

Herbert, A., Burton, C. P., & Hartley, D. (2017). The EDSAC replica project. In M. Campbell-Kelly (Ed.), *Making IT work: Proceedings* (pp. 22–34). British Computer Society.

Keramidas, K. (2015). *The interface experience: A user's guide*. Bard Graduate Center.

Link, D. (2016). *Archaeology of algorithms and artefacts*. Univocal.

Murrell, K., & Holroyd, D. (2013). *The Harwell dekatron computer*. The National Museum of Computing.

National Museum of Computing. (2022a). *Mission statement of the National Museum of Computing*. www.tnmoc.org/our-story

National Museum of Computing. (2022b). *EDSAC – electronic delay storage automatic calculator*. www.tnmoc.org/edsac

Rojas, R., & Hashagen, U. (2000). *The first computers: History and architectures*. MIT Press.

Sale, A. E. (1998). *The Colossus computer, 1943–1996*. M & M Baldwin.

Sale, A. E. (2004). *An interactive computer simulation of the World War II Colossus computer*. www.codesandciphers.org.uk/anoraks/lorenz/tools/vcoltx4.pdf

Sale, A. E. (2005). The rebuilding of colossus at Bletchley Park. *IEEE Annals of the History of Computing*, *27*(3), 61–69. https://doi.org/10.1109/MAHC.2005.47

4 Social media enters the museum

Collecting WeChat at the Victoria and Albert Museum

*Natalie Kane, Corinna Gardner,
and Juhee Park*

The Victoria and Albert Museum (V&A) in London is the UK's national museum of art, design, and performance and has engaged in contemporary collecting since its foundation as the Museum of Manufactures in 1852. Digital art and design have formed part of the collection since the late 1960s when the first examples of computer art were acquired following the *Cybernetic Serendipity* exhibition held at the Institute of Contemporary Arts in London in 1968. New additions to the collection followed in continued response to the changing technological and creative landscapes of the twentieth and twenty-first centuries.

Today digital design at the museum enfranchises a broad range of objects and practices, including but not limited to, video games, app design, mobile phones, and emojis. Digital design forms a key part of contemporary design practice. It has therefore become key for the institution's collecting policies to accommodate the shifting nature of the industry and of practitioners and users today. Through the examination of the collection in one particular case, the 2017 acquisition of WeChat, this chapter seeks to outline the ways in which digital design has manifested within a museum that seeks to collect design. In doing so, this chapter addresses the specific curatorial challenges faced in collecting objects of such complexity but also points towards subsequent curatorial opportunities as the museum's collections research and policies have developed. As the chapter will show, the case of WeChat is important not only because it demonstrates the first time that a major collecting institution has acquired a social network application as a design piece but also because it provides insights into both the challenges and the potential for innovation that developing practices to collect and curate digital design entails.

The context: digital design and society at the V&A

The V&A's 2019 Collections Development Policy outlines the collecting ambitions of the Design Architecture and Digital Department (DAD). This

DOI: 10.4324/9781003424703-8

curatorial department was established in 2015, emerging from the Contemporary Architecture, Design, and Digital Department (CADD), which was first created at the museum in 2013. DAD had set out a clear mandate to engage with contemporary trends, and the department sought to develop a collection that both expanded the definition of design and interrogated the role it played in society. Particular focus was placed on issues of manufacture, consumption, and production, while showcasing innovative and critical design practices.

Digital design has grown rapidly as a field practice and has made a clear impact on both material culture and society more broadly. As such, it has been increasingly recognised as a vital part of the museum's collecting remit. It was first identified as an area of explicit interest in 2010, and the recognition of its increasing importance is reflected in the establishment of a team and then a department with a specific mandate to collect digital design. In 2015, CADD's remit was to "collect the work of architects, designers and digital designers whose practice addresses themes of public life"; by 2019, DAD's collecting priorities had been extended to include an emphasis on design and society (Long, 2013). The focus on design and society enabled a new direction for digital collecting, placing emphasis on the impact of networked technologies on design and the influence of the internet on the way in which design is "produced, applied and disseminated" (Victoria and Albert Museum, 2019). It also introduced an impetus by which the collection would aim to make the intangible nature of digital culture visible through the collection of critical artists and designers. Continuing the DAD department's emphasis on the social role of design, this new strategy sought to represent how the "rise of design that is born digitally or mediated through digital means has had a significant impact on public life and culture," and the department sought to acquire objects that reflected these ambitions (Victoria and Albert Museum, 2019). As a future collecting priority, a remit for the collection of tools of digital design was also identified, such as those for computer graphics and animation like Blender or Cinema 4D, or as a closer study of the design process of three-dimensional printing both through hardware and three-dimensional data. This remit also constituted a commitment to expanding the V&A's understanding of digital process.

This clear set of strategies marked the first time that the museum had set out a direction for the collection of digital design and included an attempt to define the field and its boundaries. Inevitably, this led to a significant growth in the museum's holdings of born-digital and hybrid digital design, expanding in recent years from hardware such as smartphones and smart devices that has previously been held within the collection to represent design and technology, to more complex born-digital and hybrid objects – that is to say, objects with both digital and physical component parts – including apps, software, and web-based videogames in recent years. Alongside the definition of a clear collecting strategy, digital design collection has also been shaped by CADD's instigation and ownership of the Rapid Response Collecting program. Launched in 2014, the initiative aimed to ensure that V&A collecting practices "reveal the reality of contemporary design and manufacturing for future researchers by acquiring objects in timely response to global events" (Long & Gardner, 2014). Several digital design objects have entered the collection as part of the Rapid Response Collecting program, ranging from the smartphone game Flappy Bird, the X-TIGI mobile phone and battery charger, to the mosquito emoji designed by Aphelandra Messer. Within the remit of the Design and Digital department, there are now approximately 400 born-digital and hybrid objects in the V&A collection.

Collecting WeChat at the V&A

The acquisition of WeChat (微信; Weixin) in 2017 was the first time that the V&A, or any major collecting institution had acquired a social network as an item of design. The Library of Congress had previously attempted a project to preserve on a large scale an archive of social media site Twitter, collecting four years of data from the site's creation in 2006 up until 2010. However, due to the change in scale and use of the platform, continuing the project became unviable, and the Library decided to collect tweets on a more selective basis (Osterberg, 2017). The Museum of London had acquired a number of tweets in. csv format as part of their #citizenscurator project for the 2012 London Olympics (Ride, 2013). The Collecting Social Photo project by Aalborg City Archives, the Finnish Museum of Photography, the Nordiska Museet, and the Stockholm County Museum focused on the collection of memes and born-digital social photography through FINISH (Rees, 2021). However, all of the previous projects prioritized social media data rather than the preservation of the platform or user experience design itself.

WeChat was released in China on 21 January 2011 by Shenzhen-based media and technology company Tencent. Also known as Weixin (which translates as 'micro-messages'), the social network featured audio messages as a central function. Over the years it has expanded to include many more functions, so much so that it became known as many apps within one app. The app became so omnipresent in China that doing anything without it presented a challenge – cash transactions, ordering a cab, and even accessing public transport through QR codes all happened via WeChat. Within China, so much of social and administrative life is now conducted through the app that it is increasingly difficult to find alternatives to perform the same tasks. WeChat was designed to be "super-sticky" (Chen, Mao and Qiu, 2018), ensuring users adhere to it as a fundamental feature of their everyday life (Cormier, 2016). At the time of acquisition in 2017, according to Tencent's design team, WeChat had 570 million active users.

The acquisition of WeChat was proposed in 2015 by V&A curators Luisa Mengoni, Corinna Gardner, and Brendan Cormier, who recognized the all-encompassing design ambition of the app, its significance for the future of social media design globally, as well as the social impact it has had on its millions of users. Luisa Mengoni had experienced the usefulness of the app personally as Head of the V&A Gallery at Design Society in Shekou. It proved an invaluable tool for collaborative and organizational tasks, "an absolute requirement . . . of daily working life in China," as Brendan Cormier (2017) puts it.

WeChat was designed using principles of 'careful design' and simplicity, insofar as the designers fused traditional Chinese design theories of white space together with human-oriented Japanese graphic and product design principles, to create a sense of 'space and balance' for the interface and user experience of the platform (Cormier, 2016). The comprehensiveness and complexity of the platform's functional design – an attempt to "connect everything that can be connected" between user, device, and enterprise, as emphasized by Kink Weng, Head of Design for WeChat, in an interview conducted by the V&A in 2016 at the time of acquisition (Cormier, 2016) – presented an important aspect of the app to capture, both in terms of the objects collected and in the accompanying collections management and conservation documentation. Inevitably, this too became part of the challenge from the perspective of digital preservation.

Curatorial autonomy and corporate collecting

The V&A's acquisition was built on the relationship fostered with the Tencent WeChat design team over a two-year period, starting with an initial visit to the company's Guangzhou campus in July 2015. At the time of acquisition, a series of items were chosen to represent the design of this complex platform. These include a 2011 demo. apk file of the first release of WeChat, which captures the inception of the platform at the point of its creation and which was used to pitch the network to Apple's AppStore. This was provided on both a mobile handset and in digital form for storage on the V&A's digital asset management system. A 2017. apk file was acquired to mark the five year anniversary of the application, alongside 150 Bubble Pup. gifs and a series of drawings on paper that detailed the conception of Bubble Pup's. gif design and the storyboarding used by the Tencent designers. Bubble Pup was the 'unofficial WeChat mascot' at the time.

It is important to again emphasize that this acquisition was made possible due to the relationship brokered by Luisa Mengoni and Brendan Cormier with WeChat's parent company Tencent from 2015 onwards. This was facilitated by the curators' proximity to the company's campus, as the team worked on the Design Society project at this point, which aided long-term communication and the ability to reach across language barriers and the use of translation services. This relationship raised a number of significant curatorial challenges for the process of acquisition. Although the curatorial team were afforded privileged access to exclusive material – including hand-drawn sketches and access to a version of the 2011 original version of the app which would enable potentially invaluable options for digital preservation due to its availability offline, as well as access to the company during the acquisition for interview and clarification – this access was not without trade-offs. For the 2017 version of the app, a copy was loaded onto a handset and copied onto the V&A's digital asset management system, and a profile was created for a fictional personality named 'Star,' created entirely by Tencent, in order to demonstrate the social features of the application. This was meant to circumvent the multiple intellectual property, privacy, and copyright issues that accompany the acquisition of 'real' user-generated content (and the content of those people a user would interact with), both at the point of acquisition and into the future. The issue of user-generated content and social media platforms, particularly in preserving the experience or context for future audiences and for future study, is in fact a significant challenge for digital preservation as identified by the Digital Preservation Coalition (Thomson, 2016). This is due to the privacy, data protection, and ethical complications of institutional archives and collections holding personally identifiable data within their holdings.

In terms of representing the vitality of WeChat's life as a social network, arguably Star is not a representative reflection of WeChat's user base and how WeChat is used and is perhaps closer to a characterization of Tencent's ideal user. This presents potential problems in representing the use of the platform for future scholars of social media design. This issue of curatorial autonomy when working with large partners to gain access to digital cultural heritage is in growing need of its own scholarly research focus, particularly in trying to understand the lines that are negotiated and the agency and roles museums and cultural institutions hold in representing this material in context for audiences, without significant corporate intervention (Arrigoni et al., 2022). For the V&A, this was a necessary compromise. Although Star presents a slightly distorted example of one aspect of the platform's design, it nonetheless opened up the potential for interpretation in the future (within the constraints of digital preservation) where perhaps none would otherwise exist.

WeChat and collecting auxiliary objects

Interestingly, the WeChat application was not the first product from the WeChat family to enter the V&A. The initial object to enter the collection that spoke to the life of WeChat was a small, pink dinosaur-type creature called 'Mon Mon,' a soft toy that connected to the platform and enabled users to send voice messages and emotive instructions to each other. It had been exhibited previously as part of the Rapid Response Collecting program in 2015, by senior curator of design and digital, Corinna Gardner, and for *Unidentified Acts of Design* in Shenzhen in 2015 by curators Sunny Cheung and Brendan Cormier. This object was acquired independently of the larger acquisition with Tencent and spoke more directly to a specific part of the social context that frames the use of WeChat, that is to say, one of distance and longing across families when parents work away from home for long periods of time. Mon Mon didn't need WeChat's app platform to function for audiences as it was not turned on or indeed connected to the network as part of the acquisition process, or for display purposes. Its form and packaging alone were considered to stand as evidence of the social life of its users beyond the screen.

When the WeChat app entered the collection and was put on public display at the V&A in South Kensington in 2017, Brendan Cormier expressed the issue of experience and distance with contemporary digital objects such as WeChat, emphasizing that a static digital platform on a phone would be uninteresting, especially for typical WeChat users (Cormier, 2017). Therefore, a demonstration video was created by Luisa Mengoni and the Tencent team, in order to walk users through the app's variety of functions, with subtitles provided in both Chinese and English. This was storyboarded and scripted by Tencent and the V&A's curators in the DAD department, to create an experience that would explain the various functions to those who were unfamiliar with the platform. In addition, this video acted as a record of design function, capturing, with some fidelity, the layers of use that an individual and their community may encounter on the WeChat platform.

The display of the video was first located within the V&A's Twentieth-Century gallery in 2017, where it was placed next to a selection of the Bubble Pup .gifs and the hand-drawn sketches that had been acquired (Figure 4.1). The looped video, which was screened on a Samsung mobile handset, served to demonstrate the object in use. For many of our objects, putting them on display is the step that follows acquisition, allowing us to understand their place in people's lives. For an object that is still so active in the lives of many people in both East- and Southeast Asia and across the globe, to encounter it in a gallery setting is potentially a critical moment of estrangement from something that previously felt familiar. This is often the case when exhibiting contemporary design artefacts, such as the iPhone 6 that was put on display in 2016. This object is familiar to many people in a deeply embodied way; to see it within the museum frame and with attention brought to issues of scale, mass-manufacture, and consumerism, allowed audiences to reflect on their relationship with a vast network that can potentially seem intangible to grasp. How does it feel to be a part of digital design history as it happens?

Most recently, both the video and another selection of the Bubble Pup. gifs went on display in 2021 in the new Design 1900–Now gallery. They were arranged in a section of the gallery focusing on the impact of data and communication on design and society from 2000 onwards. In part prompted

Figure 4.1 WeChat demonstration video, Bubble Pup GIFs and sketches in Gallery 76, Victoria & Albert Museum.

Source: Photo by Peter Kelleher

by this renewed engagement with the set of WeChat components, curators Corinna Gardner and Natalie Kane from the department's Design and Digital Section made the decision to acquire the demonstration video of WeChat as an object in its own right. This retrospective decision made in 2022 was made on the basis of a long-term understanding of the object's life in the gallery and its relationship with audiences that experienced it, of the changing status of the object's auxiliary items, and as a form of documentation of objects in digital preservation.

Through tours that focused on the object and through further interaction with its life in the gallery, many visitors who had used the object daily and then encountered it in the museum space often found the display object a means by which to reflect on and share their own experience of using the platform. Through the detailing of the multiple capabilities of the design of the object, as opposed to a more static or 'contained' artefact or singular user experience, the exhibition item became a conversation starter. Our audiences had become an important part of redefining the object's history through their interaction with it during display, which in turn becomes part of its preservation as a complex digital object (Arrigoni et al., 2022). The difficulty since the beginning of the process of collecting WeChat had been in representing a networked, 'live' digital object, which like other social media platforms might be better understood as a form of complex assemblages (Zuanni, 2021).

Therefore, the decision to collect an artefact that resonates with the experience of networked use while capturing the broad reach of its design felt like an appropriate decision in adding to WeChat's representation in the collection. Doing so retrospectively reflects the importance of adopting experimental approaches to digital collecting. This can require more flexible policies and procedures to allow the digital objects to evolve over time. This was a key recommendation from the Towards a National Collection discovery project report *Preserving and Sharing Born-Digital Objects Across the National Collection* (Arrigoni et al., 2022).

From demonstration to futureproofing

At the point of creation, the video was conceived as an important tool of demonstration, rather than an item of documentation and futureproofing. However, over a short period of time, it has proved to be an important record of design that will likely outlive the ability to keep and/or resurrect the applications currently held digitally, whether virtually or on handsets, in long-term storage on the V&A's digital asset management system (DAMS). There are multiple challenges that accompany the collection of social media, as is evident in the history of preserving net art (Dekker, 2018), whereby the connection to a platform and proprietary network provides a unique problem for conservation, as removal from the native platform creates multiple issues in the recreation of the original experience for future audiences. There are also significant issues relating to the emulation and preservation of mobile applications, with limited scholarship existing on their long-term care (Van der Kniff, 2021; Pennock et al., 2019).

The WeChat video documentation is now recognized as a record of the design of a platform in use at the time of acquisition. Furthermore, it offers a vital snapshot, both as a means of providing context for those who are unfamiliar with the design, and as a brief portrait of social life on the application. However, authorship of the video is important to consider, as no documentation is neutral. For example, the WeChat video emerged from a collaboration from Tencent, and the narrative is very much influenced by their interventions. Similarly, the decision as to who is chosen to create these videos and provide specific windows onto the design has significant consequences. To give an example, a university student's experience on Facebook during the 2010s would be markedly different to that of a 50-year-old's today. Whether for a user, a designer, or otherwise, the record of a digital platform constitutes evidence of its existence, and both a record of social history and conservation practice – albeit one that is part of the "messy practices of managing and making do" (DeSilvey, 2017, p. 22) – that often occurs as part of digital object conservation, as it constantly evolves through a practice that is fighting decay and unavoidable technological failure.

From the perspective of a design museum that seeks to capture the design of an object, it is increasingly important to capture what Samuel Alberti et al. (2018) call 'context of use' through a form of documentation that holds an auxiliary role alongside an object. Taking an experimental approach to digital collecting that sees the relationship between objects and their documentation in more fluid, relational ways that account for digital infrastructural instability (both in terms of institutional capabilities and anticipated technological obsolescence) reduces risk and opens up the potential for situating a digital artefact within a rich variety of narratives.

The V&A has taken further steps to explore other collecting processes around apps, for example with the recent acquisition of the sexual health mobile application Euki, whereby an. apk file of the platform has been acquired with a documentation video to detail its privacy-first design. This video was produced by Ibis Reproductive Health and their community creators, who continue to advise and provide consultation on the platform. This process may also trigger further reassessment of the status of documentation of other application objects that have been acquired by the museum, such as mobile game Flappy Bird, for which two documentation/conservation records exist of the application's gameplay.

Conclusion

Like many institutions approaching the field with enthusiasm, the V&A are operating during a nascent period of collecting networked complex born-digital objects. Therefore, new and rarely standardized approaches are often taken with each object – a process of research-through-acquisition. As demonstrated in this case study, digital objects pass through many hands before they reach institutions. Their relationship to corporate bodies create new tensions and relationships for collecting institutions to negotiate both in enabling an object to be displayed in the future and made available for further study and in ensuring a legible and meaningful object that users recognize and feel represent their lived experience of digital design. Therefore, there is a need for experimental approaches, both in terms of institutional policy and practice, which provide opportunities for rich, multifaceted acquisitions. There is also a need to take into account the inevitable digital preservation risks and instabilities, while simultaneously creating the opportunities to invite external collaboration, to look beyond the museum space in informal yet crucial communities of stewardship – what Annet Dekker (2018) describes as 'networks of care' – to further our understanding of the shifting and evolving nature of the digital object in society. This shows the importance of further research into born-digital collecting, and the opportunities offered by funded collaborative projects, in order to ensure that digital heritage and digital culture do not continue to become a web of fragile parts.

References

Alberti, S. J. M. M., Cox, E., Phillipson, T., & Taubman, A. (2018). Collecting contemporary science, technology and medicine. *Museum Management and Curatorship, 33*(5), 402–427. https://doi.org/10.1080/09647775.2018.1496353

Arrigoni, G., Kane, N., McKim, J., & McConnachie, S. (2022). *Preserving and sharing born-digital and hybrid objects from and across the national collection.* https://vanda-production-assets.s3.amazonaws.com/2022/01/20/12/49/45/92b733d4-929e-429e-9fd1-82d134405465/VA-ResearchReport-Jan22.pdf

Chen, Y., Mao, Z., & Qiu, J. L. (2018). *Super-sticky WeChat and chinese society.* Emerald Publishing.

Cormier, B. (2016). *UAOD part 6: The platform/WeChat.* www.vam.ac.uk/blog/international-initiatives/uaod-part-6-the-platform-wechat

Cormier, B. (2017). *How we collected WeChat.* www.vam.ac.uk/blog/international-initiatives/how-we-collected-wechat

Dekker, A. (2018). *Collecting and conserving net art: Moving beyond conventional methods.* Routledge.

DeSilvey, C. (2017). *Curated decay: Heritage beyond saving.* University of Minnesota Press.

Long, K. (2013). Opinion: Kieran Long on 95 Theses for contemporary museum curation. [online] Available at: https://www.dezeen.com/2013/09/12/opinion-kieran-long-on-contemporary-museum-curation/ [Accessed 4 Jun. 2024]. Dezeen.

Osterberg, G. (2017). *Update on the twitter archive at the library of congress.* https://blogs.loc.gov/loc/2017/12/update-on-the-twitter-archive-at-the-library-of-congress-2

Pennock, M., May, P., & Day, M. (2019). *Considerations on the acquisition and preservation of mobile ebook apps.* https://zenodo.org/record/3460450#.ZDlnOuxKgdU

Rees, A. J. (2021). Collecting online memetic cultures: How tho. *Museum and Society, 19*(2), 199–219. https://doi.org/10.29311/mas.v19i2.3445

Ride, P. (2013). Creating #citizencurators: Putting twitter into museum showcase. In K. Cleland, L. Fisher, & R. Harley (Eds.), *Proceedings of the 19th international symposium of electronic art, ISEA2013.* University of Sydney.

Thomson, S. (2016, February). *Preserving social media. DPC technology watch report.* www.dpconline.org/docs/technology-watch-reports/1486-twr16-01/file

Van der Kniff, J. (2021). *Towards a preservation workflow for mobile apps.* www.bitsgalore.org/2021/02/24/towards-a-preservation-workflow-for-mobile-apps

Victoria and Albert Museum. (2019). *Collections development policy.* https://vanda-production-assets.s3.amazonaws.com/2019/08/05/10/18/38/9432e948-aede-479a-bdee-3cb18be965c2/2019%20Collections%20Development%20Policy.pdf

Zuanni, C. (2021). Theorizing born digital objects: Museums and contemporary materialities. *Museum and Society, 19*(2), 184–198. https://doi.org/10.29311/mas.v19i2.3790

Provocation no. 2: all of this belongs to us

Andrea Lipps

Contemporary life is mediated not only by physical matter but increasingly by digital ecosystems. Our virtual lives are enmeshed in an ecology of technologies just as our 'analogue' existence is entangled in a web of material artefacts. The exponential growth and interdependencies of both hardware and software have dramatically reshaped human life, and we in turn redesign those media in an infinite feedback loop (Colomina & Wigley, 2016). As institutions of memory, museums collect, steward, display, and interpret objects to narrate these histories as part of our shared cultural heritage (Parry, 2013).

Since 1897 Cooper Hewitt, Smithsonian Design Museum in New York—America's design museum—has collected physical design artefacts that shape the meaning and function of human life. The museum accessioned its first born-digital work in 2011 and slowly built a young collection of born-digital design until, in 2023, the museum formally established the Digital curatorial department (Cooper Hewitt, Smithsonian Design Museum, 2023). The Digital is the first new collecting department in the museum's 125-year history and joins its four existing departments—Drawings, Prints and Graphic Design; Product Design and Decorative Arts; Textiles; and Wallcoverings. This milestone signals the museum's recognition of not only the radical shift in the role that digital technologies play in our world but the importance of treating the digital as a cultural object worthy of museum acquisition and stewardship.

To be clear, digital processes already touch many areas of contemporary design practice that cut across both Cooper Hewitt's and many other museums' existing physical collections. As a tool, digital technologies have transformed the design process since the late twentieth century, yielding new methods in the drawing, planning, and fabricating of material work such as consumer products, electronics, objects, jewelry, posters, textiles, wearable technology, architecture, and more.

But in consideration of life inscribed *on* computer technology and not simply by it, museums capture digital objects and experiences themselves, alongside their histories, spheres of use, and display possibilities. The practice of interactions, data visualizations, apps, websites, information architecture,

DOI: 10.4324/9781003424703-9

videogames, interfaces, and more are central areas of digital design that are woven into and on which our everyday experiences are etched. For museums, collecting such digital artefacts is to acknowledge their importance within the realms of visual and material culture, by positioning them as part of the historical record. We like to think of this work as being part of the cultural commons, preserving stories for future generations about our digital lives: how they develop, shape, and in turn are redesigned by us at increasing speed.

And yet these experiences do not occur in a vacuum; they are contingent on an ecosystem of hardware dependencies, networks and libraries, increasingly robust software, and communities of stakeholders. Despite the complexity, museum collections engage with digital media because sharing histories of contemporary human experience while omitting the digital would be like publishing books while omitting the words. It would be to miss the connective tissues that bind both our material and virtual worlds. For Cooper Hewitt, a public institution of memory, all of this belongs to us. It is both physical and digital artefacts that help us understand our material histories and contemplate the trajectories that propel us, enabling reflection on what it is to be human.

Reference List

Colomina, B., & Wigley, M. (2016). *Are we human? Notes on an archaeology of design*. Lars Müller Publishers.

Cooper Hewitt, Smithsonian Design Museum. (2023). *Cooper Hewitt announces formal establishment of the digital curatorial department*. www.cooper hewitt.org/2023/03/09/cooper-hewitt-announces-formal-establishment-of-the-digital-curatorial-department/

Parry, R. (2013). The end of the beginning: Normativity in the postdigital museum. *Museum Worlds*, *1*(1), 24–39. https://doi.org/10.3167/armw.2013.010103

Part III

Living computing history collections

5 Mediators, media, and meaning

Curating digital objects at the Science Museum

Tilly Blyth and Rachel Boon

The Science Museum has a long history of collecting and displaying digital computing and data processing technologies, from demonstration models of Charles Babbage's analytical engine and some of the first digital electronic computers and programs in the world, to digital networks and mobile devices. Using examples from the museum's collections, we consider the shift from 'hero' collecting, which risks presenting scientific innovation through a selection of 'highlight' machines and key human protagonists, to using material culture as a lens through which to explore how digital technologies and data are culturally, socially, and politically embedded. In this way, the exploration of digital objects in museum collections provides a tool for understanding ourselves and offers a unique insight into the ever-changing human condition and the role of digital technologies in some of the most challenging issues of our time.

Traditional approaches to collecting punched cards, magnetic tape, and floppy disks, reflect the practice of data storage and manipulation, whilst other tangible digital technologies such as cables, routers, and servers represent our need to share and transfer data. However, should a museum collect, preserve, and present the material that places these technologies within a broader context, to provide a framework through which to understand their role and significance within society itself? What role does such material play for the organization, its audiences, and for historians and society as a whole? Can the organization deliver if it only collects and preserves the physical artefact, without the supporting data that reflects the context of use? This chapter reviews and recommends curatorial approaches to collection development and display that help to navigate the increasing intangible nature of computing and data processing technologies. In doing so, we highlight the unique challenges and opportunities raised by the proliferation of immaterial technologies and attempt to show how user-focused or embedded curatorial approaches to collecting might result in a more meaningful understanding of the role of computing technology and data in our society. We argue that to date, the development of museum collections has been very partial; that there is a need for museums to reflect a broader range of narratives, including

DOI: 10.4324/9781003424703-11

those relating to their use in a social context; and that this can only happen if museums begin to routinely acquire and preserve a broader variety of data and documentation to support artefact collections.

Creating a collection

Beginning life within the Mathematics Collection, the Science Museum's calculating machines formed a notable part of a wider collection used to represent teaching, computation, and technical drawing. These included mechanical aids for calculation, as well as machines powered by electric motors, and later electronic devices. The Mathematics Collection prioritized practical mathematics and tools for calculation as a key focus, comprising Difference and Analytical Engines, instruments for solving equations, and Harmonic Analysers and Integrators. During the interwar period the museum developed these collections to reflect changes to data management tools that emerged in the early twentieth century. This consisted of the acquisition of semi-automatic and electrical tabulating machines made by the British Tabulation Machine (BTM) Company. By the 1950s the collection presented many 'firsts' in computing history, such as Andrew Booth's magnetic drum store, the National Physical Laboratory's (NPL) Pilot ACE computer, and parts of Cambridge University's Electronic Delay Storage Automatic Calculator computer (EDSAC). Through these acquisitions, the seeds had been laid for the representation of Britain's fledgling electronic computing industry in museum collections. It is perhaps not surprising that these 'electronic brains,' as they were contentiously called (Bowden, 1953, p. v), were initially placed in the Mathematics Collection, as they were often built within university mathematics departments or by teams of mathematicians and electronics research engineers. Indeed, it was not until the implications of Alan Turing's Universal Machine began to be fully realized that the use of these machines beyond rapid automatic calculation began to be fully appreciated.

Many of these objects were initially curated to communicate a triumphal and deterministic narrative of progress. Objects were donated as symbols of prestige, as scientists, engineers, and industry professionals looked to the UK's national Science Museum as a site for validation and promotion. The NPL at first lent, and later in 1956 donated, the Pilot ACE, only a few years after it was heralded as the fastest computer in the world. The Science Museum's Keeper of Astronomy, Henry Calvert, was reported to have said, "Millions of people have read about robot brains, but few have seen one" (Daily Herald, 1956). Displaying this highly significant British computer reflects the defiant modernism of the postwar era as visual icons came to represent the nation's progress through technological achievement (Bud, 1998). Today the museum's galleries illustrate how these influential digital artefacts were shaped by their wider cultural, political, and economic context and co-constructed by a range of personnel who worked with, and were affected by, these machines (Blyth, 2014).

The tension between the physical scale of postwar computer installations and the limits of museum storage space has meant curators have had to, and still continue to, acquire parts to stand in for the whole (Swade, 2020, p. 164). The evolution of the computing collection at the Science Museum reflects the institution's priorities over time. Specific computing components such as delay lines and floppy disks were displayed to explain the technical characteristics of the larger devices with which they were used. The focus was on collecting the technical innards of the larger machine to illustrate the efficiency of processing and the storage of information. This consisted of indicative modules with valves and printed circuits boards,[1] early drum stores,[2] and delay lines that show the ingenious ways that data was saved. By 1971 the growth of computing technology could no longer be captured and contained within the Mathematics Collection. Developments in electronic technology, made possible in part by the miniaturization of components such as transistors, integrated circuits and microprocessors, meant an increasing number of acquisitions of computer and data-processing material were inappropriate for inclusion in the Mathematics and Mathematical Instruments Collection. This created a need for the formation of a new Computing and Data Processing Collection (Science Museum, 1971, p. 14). Nearly a third of the Mathematics Collection was reallocated. Perhaps surprisingly, this included items that are unquestionably mechanical in design, such as those made by Charles Babbage and Georg and Edvard Scheutz.

Displaying the collection

As is often the case, the creation of the new Computing and Data Processing Collection was driven in part by display opportunities, as it was announced in the annual report a year later that, this "new Collection . . . consists of . . . a mass of recently-acquired material collected to meet the needs of the new Gallery" (Science Museum, 1972, p. 11). The Computing Then and Now gallery divided the space both physically and conceptually between analogue and digital computing (Blyth, 2013). Opening in 1975, it utilized a range of interpretive tools, including photographs of people working with the displayed objects, a 'Computer Arts Booth' curated in collaboration with the Computer Arts Society (CAS) (Swade, 2008), and even contained a pioneering computer terminal networked to a mainframe at nearby Imperial College (Science Museum, 1975). Some displays provided further context for the objects, notably a staged '1930s Office' environment, in which punch card technology was demonstrated and used. In the main, the gallery displayed computing technology within the chronological sequence of its production, with the progression of one system seamlessly moving onto the next. Whilst the gallery presented the advance of computational tools as an effortless historical progression, the lived experience of such shifts were profound, messy, and led both to the creation of other new machines and new forms of labor and societal organization. In contrast to the narrative that presented the advance in the efficiency of computers for work purposes, the computer arts booth, situated at the exit to the gallery, highlighted the potential of computers to support artistic endeavors.

Personal machines, social devices

In 1975 the Computing Then and Now gallery was state-of-the-art, but by opening shortly before the impact of the personal computer revolution, it quickly started to appear outmoded. Rather than being exhibited, these developments were captured by curators through new acquisitions to the collection. The miniaturization and mass production of technical components increased portability and the steady democratization of access to digital technologies. The rise of personal computers and devices marked a significant change in the consumer's access to computational power. Not only did it become easier for museum staff to collect entirely complete machines, but these acquisitions reflected a wider range of user, adapter, and consumer experiences of technology. As a result, new items collected into the Computing and Data Processing Collection no longer only reflected utilitarian tools – for example, machines designed for the purpose of payroll tasks, creating census records, and performing banking calculations – driven by the speed of processing and the ability to store data, but also new devices that supported creative thought and imaginative production.

The introduction of the Graphical User Interface, on devices such as the Apple LISA and the PERQ workstation, ensured that anyone could use a computer without needing to learn to code. As computers quickly developed into machines that enabled word processing, desktop publishing, illustration, gaming, and architectural design, in contrast to older devices intended solely for rapid calculation and manipulating spreadsheets, museum curators attempted to represent computers as tools for more imaginative endeavors. Curators also broadened acquisitions to include the context of their use, from computer manuals and mice, to educational software and gaming apparatus, thus beginning to reflect the wider lived experience of the consumers that used and operated these machines.

Throughout the 1980s, computers became more portable and personal, with the Osborne 1, Psion Organiser, Apple Newton, and Palm Pilot, all adopted by the consumer to reflect social status and facilitate a lifestyle defined by the user. As these artefacts were acquired by the museum, they no longer simply represented the march of technological and social progress, measured through processing speeds or the storage capacity of bytes, but the personal experience of the user. The software they ran represented the social hierarchies of society (Cooper & Woolgar, 1993) and the form of the hardware as was understood as resulting from the social conditions of their creation. These were no longer artefacts that merely provided material evidence of technical achievement but were instead imbued with the meanings of the everyday lives of their users as they were defined and redefined through the context of their creation and use.

Objects as networks

Throughout the 1960s and 1970s, people explored connecting computers and their peripherals, using modems, cables, and timesharing devices, to create early data networks such as ARPANET and Local Area Networks (LANs), or digital information services such as Minitel and Prestel. The Science Museum represented these developments in its telecommunication and computing artefacts, acquiring acoustic coupler modems, teletype terminals, and both copper and prototype fiber optic cables. However, it was not until the use of such networks exploded with the birth of the World Wide Web in 1990 that a networked approach to the Computing and Data Processing Collection took center stage in both the museum's collections and displays. Curators at the museum acquired the personal devices that people used to connect to the internet, including computers not only such as the internet-connected iMac, the Amstrad notebook, and digital modems, but also the hidden devices that supported the infrastructure of the internet and the World Wide Web, such as network routers, switches, and servers.

Many of these objects were placed on display in the museum's landmark Information Age gallery, which opened in 2014. The gallery explores the last 200 years of information and communication networks through six networks – the telegraph, radio and television, telephone, satellite, computer, and mobile phone (Blyth, 2014). However, rather than construct a story of 'revolution' with each new technology replacing the next, the gallery presents a history of 'coexistence' between the networks. At the conceptual core of the gallery is an understanding that, whilst we may feel we are living in an Information Age, the experience of our predecessors in terms of the speed and distance of new communication networks was just as revolutionary, if not more so, as our own today. The tale is told through the stories of those who invented, made, operated, or were affected by these new waves of technology. In this way the gallery reflects an awareness that "all societies today have a relationship with computing, not only those that have 'pioneered' computing or even those labelled as early or late adopters" (Misa, 2007, p. 53). The Information Age gallery is primarily narrative-led, with many of the stories told by women and diverse communities, to encourage audiences to consider the machines not only through the lens of their predominantly white male inventors, but to also reflect upon the wider social uses, meanings, and implications of the technology. The objects displayed not only represent the development and use of *technological* networks, but the people, places, and ideas that were *socially* entangled with these networks.

Displaying and acquiring data

Whilst the Information Age gallery is rich in artefacts, it also explores the immaterial aspect of the data that is shared through networks. Each of the six networks are signposted by a large black 'Story Box.' The media shared through these networks are subject to their own form of interpretation within the gallery. For example, 'The Broadcast' Story Box explores the most listened to or watched programs in history of UK radio and television. At the 'The Web' Story Box, Sir Tim Berners-Lee addresses the question of 'what happens when you click on a link?', exploring the complexity of information sharing through web addresses, hypertext markup language (HTML), and packet switching, all explained by the very individual who invented the World Wide Web. 'The Cell' Story Box explores the cellular networks underpinning our mobile devices. A series of interlinked, character-led narratives illustrate contemporary 'digital life,' showing visitors how cellular networks enable users to stay in constant contact with one another, therefore providing new opportunities for communication between diverse global communities. Visitors are invited to interact with the digital installation via their mobile phone to unlock enhanced content relating to the narratives projected around them as well as contribute their own messages, images, and thoughts that are fed into an anonymized live data record of the gallery's visitors.

It can be difficult to comprehend the power of data processing when most people's experience of data takes place through a personal device that is experienced as discrete from the wider infrastructure that supports it. This challenge was explored in the Science Museum's temporary exhibition *Our Lives in Data* (2016–2017), which argued that we are constantly producing data, that our own data can be compared with others to find patterns, and that these patterns inform decisions made in the wider world. Visitors were shown different examples of how 'big data' shapes their lives, from influencing transport systems of the future to the transformation of genomic research. Facebook was used as an illustration of how big data is changing the way private companies function. The company lent and later donated a data center server rack that represented the physical infrastructure of the internet, and the intensive use of energy that reflects; just a small component from the six enormous data centers that house the company's 1.6 billion user profiles.

While technical objects provide a snapshot of the physical infrastructure data travels through as it is processed, the Science Museum has also developed tools to reflect the intangibility of data through ephemeral experiences and displays. The audiovisual art installation *Listening Post*, a 2005 collaboration between artist Ben Ruben and statistician Mark Hansen, was put on display at the Science Museum between 2008 and 2013 and has now entered the permanent collection. Entry into the installation produces a mesmerizing experience, resembling a contemporary 'cathedral' to the internet as it visualizes live conversations from public internet chat rooms and bulletins. Data mining

Figure 5.1 The Web Story Box uses a hypertext narrative and live data feeds to explain some fundamental aspects of the World Wide Web.

Source: Courtesy of Science Museum/Science & Society Picture Library

software filters live conversations into phrases, such as 'I like' or 'I love,' which are displayed on a hanging grid of 231 LED screens. A text-to-speech synthesizer voices some of these spoken phrases and animates them within an emotive soundtrack. *Listening Post* gives visitors an unfiltered, powerful insight into people's personal data – their thoughts, their likes, and their dislikes – while raising larger questions about the notion of private and public space in the Digital Age. Whilst the display doesn't attempt to store the data and will only be able to run if such data live feeds continue to exist, it provides an insight into the experiences and limitations of human communication and reflects the vitality of seemingly mundane data relating to personal exchanges in our everyday lives.

These examples demonstrate the ways of displaying data and acquiring the material culture that supports data exchange. However, they do not address the very real dilemma of acquiring and preserving data in the longer term, with all of the copyright, conservation, and cost issues that this involves. Born-digital material exists on a network of distributed data centers, social media platforms, and personal devices, much of which is 'black boxed.' They facilitate the seamless flow of information and data between these components and the human agents who share and curate user-generated content. While we have noted that the perceived intangibility of computing culture is in some sense material, reliant on people, places, networked devices, and infrastructure, museums still face a huge challenge in deciding what it means to acquire and preserve a born-digital object in the long term and what that means for the nature of a museum in the digital future.

The future of collecting data

Although it is not necessary to collect the data passing between material artefacts to give meaning to the technology of such digital and physical objects, curators are already branching out to engage on a deeper level with the contexts of physical and digital object development and use, advocating the preservation of computing and data processing in its broadest sense, as a form of heritage for future research and display. As Boyle and Cliff (2014) have shown, science museum curators already focus on the 'people and processes' of scientific practice, rather than acquiring items because they are sensational in physical scale by today's standards (the Large Hadron Collider, for example) or completely digital (for example, datasets used at the CERN, the European Organisation for Nuclear Research). Instead, they have focused on acquiring items that are representative of working styles and events, rather than the instruments or products of science.

Curators also support physical acquisitions of Computing and Data Processing artefacts that are accompanied with contextual material, ranging from academic papers that use computing devices, promotional and marketing material released by the computing industry, and the testimonies of people that have used the machines. To engage further with the meaning of these technologies and their significance within the wider world of human experience, such an approach could be extended, with curators embedding themselves in a network of developers, marketeers, or users at the time of acquisition, so that they might better replicate in museum holdings the working lives, tools, and environments in which digital technologies are made and used. If collecting took place through highly active forms of collaboration with technological communities – and developers and users provided access to people and spaces – then a larger range of experiences and perspectives might be reflected in the documentation and interpretation of computer collections preserved in museums. This would also offer a greater sense of equity and fairness in the representation of museum collections. Curators would reflect the process of meaning-making around physical and born-digital objects across a wider range of communities, rather than acquiring the products of technological invention and use according to their own siloed perspectives. In doing so, the role of museums and their collections would shift from the didactic 'transmission' of cultural capital to audiences, towards, as Michael John Gorman suggests (2020, p. 137), facilitating audiences' access to social capital, therefore enhancing opportunities for understanding and cohesion in society.

As well as collecting contextual material for physical artefacts, we must surely adopt a similar strategy for born-digital objects. Rather than only collecting the 'real thing,' a wide range of digital and physical contextual material could be brought into use, such as. pdfs of marketing material,. jpeg screengrabs of applications, documents or audio recordings that describe user experience, videos of the object in use, or recordings of a phone screen showing the use of an application. This would create a 'curated digital artefact,' a discrete category of museum object based on an assemblage of material.

At the Science Museum, traditionally this type of contextual material has been stored physically in technical files associated with an object. Such material has in the past been understood as peripheral rather than central to the acquisition. Unlike the objects they represent, these files and their contents are not shared publicly but can be accessed on request. Such content can be fragmentary, with the level of detail obtained on the wider context of the items they acquire dependent on the enthusiastic contributions of past curators or individuals external to the museum. However, it may now be time to privilege this contextual content and to create new digital files that support knowledge and understanding around a born-digital object for decades to come. Either way, it is clearly important that any curated digital artefact needs to receive the same careful conservation and preservation treatment as physical objects in collections, including a cohesive digital preservation strategy and a clear plan to assist public accessibility. Unfortunately, in many organizations these rigorous processes for the collection and interpretation of digital objects are not yet in place. This raises a significant question about the future of museum collections, as they fail to acknowledge that nearly all of our technological systems are now hybrid, and to preserve the physical artefact without a digital counterpart constitutes at best only a partial understanding.

Conclusion

This chapter has explored how one national science museum's Computing and Data Processing collections have been formed, theorized, collected, and displayed since the nineteenth century, and the integral role data plays in the history of computing at the Science Museum. Over the last 50 years the Science Museum has moved away from focusing solely on the development of technology and how computers use data to operate, towards a contextualization of the role of data in shaping lived experience within society. Material culture sits at the core of all these stories, illustrating the physical developments in technology, and representing the people – such as the developers, manufacturers, maintainers, users, and adapters – at the center of these changes. Punched cards, data servers, and mobile phones have provided a gateway into stories about gendered labor in the pre-digital computer age, globalization, the environmental impact of data processing infrastructure, as well as individual rights in the debates about data ownership.

Capturing and representing these diverse accounts in museum collections is key to making the role of data in the history of computing relevant to our audiences. While acquiring and preserving raw data comes with both practical and intellectual challenges, the case studies illustrate that it is possible to communicate a range of narratives around the development and use of technology without access to, or preservation of, supporting data. Computing technologies are ubiquitous in the everyday experience of billions of people around the world. Histories of computing play an important part not only in our technological heritage but our cultural heritage too. Science museums have the potential to play a key role, no longer as an unremitting authority in illustrating the technological proficiency of humanity, and in particular Western societies, but as a partner in dialogue with its audiences, exploring how communities negotiate each new wave of technological change. Museums can be a key protagonist, providing the historical context for these changes, facilitating the emergence of a deeper and more diverse understanding of how computing has contributed to shape the past, the present, and the future of our societies.

Notes

1 See for example the logic door from the Ferranti Mark 1 computer, 1951, or module from the American Semi-Automatic Ground Environment (SAGE), 1963.
2 See the magnetic drum store from the Deuce computer, 1955.

Reference List

Blyth, T. (2013). Narratives in the history of computing: Constructing the information age gallery at the science museum. In A. Tatnall, T. Blyth, & R. Johnson (Eds.), *IFIP international conference on the history of computing* (pp. 25–34). Springer.

Blyth, T. (2014). *Information age: Six networks that changed our world.* Scala Arts & Heritage.

Bowden, B. V. (1953). Foreword. In B. V. Bowden (Ed.), *Faster than thought: A symposium on digital computing machines* (p. v). Sir Isaac Pitman & Sons.

Boyle, A., & Cliff, H. (2014). Curating the collider: Using place to engage museum visitors with particle physics, *Science Museum Group Journal, 2*(2). https://doi.org/10.15180/140207/001

Bud, R. (1998). Penicillin and the new elizabethans. *The British Journal for the History of Science, 31*(3), 305–333. https://doi.org/10/1017/S0007087498003318

Cooper, G., & Woolgar, S. (1993). *Software is society made malleable: The importance of conceptions of audience in software and research practice, PICT policy research paper no. 25.* PICT, Brunel University.

Daily Herald. (1956, June 28). The first robot gets the sack. *Daily Herald.*

Gorman, M. J. (2020). *Idea colliders: The future of science museums.* MIT Press.

Misa, T. (2007). Understanding 'how computers changed the world'. *IEEE Annals of the History of Computing, 29*(4), 52–56. https://doi.org/10.5555/1339814.1339836

Science Museum. (1971). *Science museum report for the years 1963–1972.* Science Museum.

Science Museum. (1972). *Science museum report for the years 1963–1972.* Science Museum.

Science Museum. (1975). *A guide to computing then and now.* HMSO.

Swade, D. (2008). Two cultures: Computer art at the science museum. In P. Brown, C. Gere, N. Lambert, & C. Mason (Eds.), *White heat cold logic: British computer art 1960–1980* (pp. 203–218). MIT Press.

Swade, D. (2020). Preserving the legacy of IT innovation. In L. Strous, R. Johnson, & D. Swade (Eds.), *Unimagined futures – ICT opportunities and challenges* (pp. 162–176). Springer.

6 Unsettling the narrative

Quantum computing in museum environments

Petrina Foti

Everything that we have observable knowledge of or can experience directly, from the cosmic to the microscopic, operates with the same set of principles. At the atomic level, the rules change dramatically. Chris Ferrie and Geraint Lewis referred to this discovery as one of the biggest surprises in the history of science and further explain that "when we look deep into the world of the very small, we find an unfamiliar world seething with the unpredictable and counterintuitive, in many ways very different from the functions of the cosmos" (Ferrie & Lewis, 2021, p. xiv). The development of quantum mechanics transformed our understanding of the universe and revolutionized scientific progress, from physics to chemistry to information science (Jaeger, 2018). One of the most disconcerting new avenues of exploration has been the quest to build a reliable, scalable *quantum computer* – a computer that utilizes specialized hardware to employ quantum behavior. For over 50 years, Moore's Law has proven true (Brock, 2006), although it has always been understood that at some point it would be physically impossible for transistors to be made any smaller. Quantum computing is poised to subvert this long-held principle.

Given the rapid development of classical computers and the way that society has been transformed as a consequence, it is understandable to wonder what might be achieved with a computational tool far beyond our current capacities. It should be noted that quantum computing does not exist in isolation from classical computing. Rather, it should be considered a new path of exploration whose full potential is still unknown. What new discoveries might be made? What new technology might be built? It is reasonable to ask, then, what these advances in computing technology might mean for the museum. In other words, when the essence of computing changes, what consequences might there be as to how museums will narrate the history of quantum computing? Even though quantum computing technology has not reached its full potential, it already poses significant difficulties for the museum in terms of both collection and exhibition practices. Sandra Dudley has argued that things, particularly objects in a museum, "are not what we come to know and represent: they are both more and different; we do not and cannot access all their potentialities and qualities; and we interpret, in forming our representations,

DOI: 10.4324/9781003424703-12

only those qualities that we do access" (Dudley, 2021, p. 18). With quantum technology there is no risk of an ontological misconception. The field is too new; our understanding is too limited. The greatest challenge that museums face when narrating quantum computing is how much we do not and might never know, and how little we can represent.

There has long been discussion about how the history of computer technology might best be recorded in a museum setting (Tatnall et al., 2013). Throughout the course of this volume, our authors, most of whom are themselves museum practitioners, have explored the various ways in which museums can successfully collect and exhibit classical computers and associated digital technology. Yet when faced with quantum computing, not all of these solutions will be able to be translated to accommodate the particulars of quantum systems. Therefore, this chapter will examine both what quantum computing is, how the museum has begun to record its development, and what the quest to understand this potentially revolutionary technology reveals about all computer technology.

What is quantum computing?

Although many important milestones have been reached in the past decade, at this moment in time, the potential of quantum computing to surpass classical computing in everyday use remains an objective still waiting to be achieved rather than an unarguable certainty. Yet even at this relatively early stage, the development of quantum computers opens a number of possibilities in terms of science and technology. As the Computer History Museum has framed these possibilities:

> Imagine an entirely new kind of computer based on the surprising physics of the quantum world. They leave digital computing behind in order to solve some incredibly complex calculations in a flash. Such a machine could leave our most powerful digital supercomputers behind for modeling everything from molecules to the climate, changing the way we understand our planet, develop medicines, crack codes, and analyze financial markets. This is the vision for quantum computing.
>
> (Computer History Museum, 2018)

While it is easy to understand how this vision for the future could potentially be transformative, in many ways it is a case of the unknown promises quantum computing might offer that truly spark the imagination.

Quantum computing uses quantum mechanics as the foundation for its arithmetic or logical operations. The most basic unit in classical computers is the *bit*, a binary digit, usually represented as either a value of 1 or 0. Its equivalent in quantum computing is the *qubit*, or quantum bit. Like the classical bit, the qubit is a two-state system. The difference is that the qubit can represent not only one of the two binary states but *all of its combinations at the same time*. The ability to be in a combination of states is called *superposition*. In a quantum computing system, qubits must be interconnected, which is referred to as being *entangled* (Grumbling & Horowitz, 2019, p. 2). This is the foundation that allows for quantum computing's potentially transformative processing power.

To understand this technology requires an understanding of quantum behavior, but the quantum world is unlike anything we have previously known. Quantum mechanics is mathematics-based and cannot be visualized (de Lima Marquezino et al., 2019), which is highly problematic for the museum field, given its long association with the representation of concepts through the display of visual and material culture. Although classical computing is itself mathematics-based and cannot be visualized, except through model or analogy, the basic concept of the binary code can be easily incorporated into our lexicon. The concept of diametric opposites is familiar to even very young museum visitors, and this provides a strong understanding of some basic theoretical concepts. Near or far. Up or down. Off or on. Yes or no. One or zero. The concept of superposition is much more difficult to grasp since we have no concrete point of reference to how something might be in a combination of states.

Quantum computing in museum environments

There is ample evidence as to how curators are able to react quickly and flexibly when they seek solutions to accessioning, interpreting, and exhibiting new types of museum objects (Foti, 2018, p. 51). It therefore is not at all surprising that museums, such as the Computer History Museum in California and the Cooper Hewitt Smithsonian Design Museum in New York, have begun the process of obtaining quantum computing related objects for acquisition and display. Many museums, such as the Science Museum of Tokyo in 2022, the Museum of Science of Boston in 2019, and the Deutsches Museum based in Munich, circa 2021, have offered public programs, usually in partnership with the large computer corporations building these quantum machines. The National Museum of Emerging Science and Innovation in Tokyo has explored quantum computing within the context of their larger *Create Your Future* exhibition. The New Haven Museum in Connecticut, in collaboration with the Yale Quantum Institute, presented a temporary exhibition, *The Quantum Revolution: Handcrafted in New Haven*, which ran from 13 April until 16 September 2022. This exhibition combined physical tools and equipment with artwork by a former artist-in-residence at the Yale Quantum Institute (Yale Quantum Institute, 2022).

The interdisciplinary nature of the exhibition at the New Haven Museum, in particular, demonstrates how interest in quantum computing stretches beyond science and technology museums to include those in the art sector.

There has long been a symbiotic relationship between information technology and art, with museums playing a supporting (and occasionally leading) role in this exchange. During the 1960s and 1970s, pioneering artist Lillian Schwartz worked closely with engineers from Bell labs to explore how computers could be used to create film and artwork, stretching the limits of the still developing technology (Computer History Museum, 2021). Half a century later, a new generation of artists are working in collaboration with quantum technology companies. For example, Refik Anadol's *Quantum Memories* (2020), an artwork commissioned by the National Gallery of Victoria in Melbourne, Australia, utilized the processing power of a quantum computer prototype to create new images (Jochim, 2022). This is just one of the many ways that the development of quantum computing is mirroring the invention of the modern digital computer, so it is not surprising that similar parallels can be seen in the museum sector's response to this new technology.

Whether it is within the discipline of art, science, technology, or design, it is striking how the approaches of these museums are similar to those employed in the acquisition and display of early computer technology over 50 years ago. Tilly Blyth, reflecting on the Science Museum in London, observed that "most exhibitions on computing created during the 1970s responded to the euphoria and excitement surrounding the new computational machines of the day. Often they were developed as marketing and promotional tools, commissioned by the computing companies themselves" (Blyth, 2016, p. 4). Computer technology in the current age is ubiquitous – in our hands, our cars, and even our washing machines – and it is easy to forget that during the time of large mainframes, computers were novel and scarce. Museum acquisitions would often be solicited from the corporations that built them for the sole reason that there was no other source. It is the same reason today that we see museums partnering with corporations and institutions at the forefront of quantum computing development. Museum narratives are shaped, in part, by what materials are available (Weber, 2016). Unless quantum computing has its own version of the personal computing revolution, making quantum computing technology available to all, museums must balance their own needs in constructing historical narratives with their partner corporation's desire to protect their own proprietary information. As a result, like their early classical computing counterparts, the hardware and associated physical materials form the focus in the acquisition and display of quantum computing devices.

A binary interface

While many museums have had success in presenting computer technology in operation on the exhibition floor, this is not feasible for quantum computing at these early stages of development as these machines must run in environments just above absolute zero. What then remains is the non-operational hardware itself as a mute testament to the computing milestone. For the modern museum visitor, the specialized hardware utilized in quantum computing (Grumbling & Horowitz, 2019, pp. 24–56) would likely be more evocative than a museum case containing what would essentially appear as a black plastic block, which would be virtually indistinguishable from the smartphone the visitor would almost certainly be carrying in their pocket or bag. However, computer hardware in general is not immediately intuitive for the majority of museums visitors, and specialized quantum computing hardware can be difficult to understand even for those who may have a strong computer science background. Computer hardware on exhibit requires interpretation for all but the most technically minded audiences. Furthermore, even with their specialized hardware, quantum computers are far more than just physical machines. Given that current quantum computers have to rely on classical computing interfaces, the problems of hardware-dependent software and software-dependent hardware are still at play.

As can be seen with *Quantum Memories* (2020), the artwork by Refik Anadol commissioned by the National Gallery of Victoria, quantum computing has affiliated digital outputs and is therefore still susceptible to the "impermanence and fragility" that have become "defining conditions of the digital age" (Domínguez Rubio, 2020, p. 305). In this regard, many of the challenges posed by quantum computing have naturally inherited classical computers' duality of hardware-dependent software and software-dependent hardware, specifically how objects that exist solely in a digital format cannot easily follow the collecting precedent of material objects in the same way that three-dimensional hardware is able (Foti, 2018, p. 23). However, as has been just noted, all computer hardware, whether quantum or classical, does not fully convey its purpose without the benefit of museum interpretation and curatorial insight. Computer software must be allowed to play its integral role for computer history to successfully be narrated in a museum environment.

As classical computing technology matured, it became increasingly common for the museum to confront the reality that two different examples of computer technology might look identical, as mass-produced objects do, but have entirely different software, making them two distinctly different objects (Foti, 2018, p. 35). Given the multipurpose use of a standardized desktop or laptop computer, it is impossible to understand how a particular example of computer hardware was used without being able to access its software. Within this framing, it is understood that the software is seen as inseparable from the hardware.

The museum draws attention to a dynamic whereby "digital objects require ecologies operating according to a logic in which modern categories such as authenticity, temporality, and ownership cannot be constructed and sustained in the same way, or do not necessarily mean the same" (Domínguez Rubio, 2020, p. 322). Recall the entangled qubits of a quantum computer. Taking a step back, it is then more easily understood that while digital technology might not be interconnected at its most basic units, it is interconnected and entangled at a *systematic* level. Without the servers, systems, and components used at the time the 'digital object' was created, it can be difficult, if not impossible, for that digital object then to exist with the same fidelity to form and function as the day it was acquired (Weber, 2016). Historians of computer technology have long been confronted with this issue when they seek to restore and rebuild obsolete classical computers. Just as "the rebuild is not indistinguishable from the original machine" (Agar, 1998, p. 126), so to these restored digital objects are not the same as the ones the museum first sought to preserve. Whether quantum or classical, computer software remains maddeningly elusive.

Conclusion: the quantum future

During the moment at which this chapter was written – and perhaps, very likely, at the moment you read this chapter – all futures seem possible, or superpositioned, if you will. Quantum computing might become as ubiquitous as mobile technology is today. It might only ever be a computation tool used in limited circumstances when its rapid processing power is needed. Or the future of quantum computing might lie somewhere between those two extremes. No one future is weighted more than the others. But no matter which of all possible futures we will ultimately experience, quantum computing has already played an important role for museums, ourselves, and everything that we have ever known.

Just as quantum physics upended everything that we previously thought we knew about how the universe operates, quantum computing unsettles our understanding of what computers are and can be. Given the way that we reach for our phones the moment we wish to seek some form of information, it can be claimed that computer technology has become so intrinsically linked with our lives that it now shapes our own thought processes and worldview, with all the positive and negative connotations that might imply. In keeping with that perspective, it is perhaps not too provocative to then speculate that, as computing moves beyond a binary comprised of diametric opposites, we will, in turn, more readily accept nonbinary concepts and framings and become more comfortable when our long-held tenants and beliefs prove to be inaccurate.

Recalling Sandra Dudley's words at the start of this chapter, even the most familiar computer technology is 'both more and different' than the understanding of it that we have formed since its inception, less than a century ago. Classical computers are more than their hardware can fully represent. Their software – what we often handwave as 'the digital' – is more complex and interconnected than we normally allow ourselves to believe. With this understanding, quantum computing is the exemplar of everything that we still do not know or understand about computer technology. Perhaps, then, the greatest lesson that we can learn from quantum computing is that we must reassess what we think we know about classical computing. Rather than reflecting on the difficulties that museums face as they seek to collect and exhibit computer technology (Foti, 2018), perhaps it might better be argued that museums have been so successful at applying a number of curatorial techniques to collecting and exhibiting computer technology that the sector has been able to circumvent the fundamental nature of digital technology. As we look to the future (one with or without quantum computing technology), we must reconcile museum practices with the unsettling truth that our beloved, ubiquitous, ever-prevalent computers might never be able to conform to those long-held traditions, because – just as the quantum world is understood through a very different set of principles from the rest of the observable universe – digital technology exists within a different set of rules.

Reference List

Agar, J. (1998). Digital patina: Texts, spirit and the first computer. *History and Technology*, *15*(1–2), 121–135. https://doi.org/10.1080/073415 19808581943

Blyth, T. (2016). Exhibiting information: Developing the *information age* gallery at the science museum. *Information & Culture*, *51*(1), 1–28. https://doi.org/10.7560/IC51101

Brock, D. C. (2006). *Understanding Moore's law: Four decades of innovation*. Chemical Heritage Foundation.

Computer History Museum. (2018). *Speaker series: CHM live: Inside the transformation: Quantum questions*. https://computerhistory.org/events/quantum-questions/

Computer History Museum. (2021). *Lillian F. Schwartz, 2021 fellow: For her pioneering work at the intersection of art and computing*. https://computer-history.org/profile/lillian-f-schwartz/

de Lima Marquezino, F., Portugal, R., & Lavor, C. (2019). *A primer on quantum computing*. Springer.

Domínguez Rubio, F. (2020). *Still life: Ecologies of the modern imagination at the art museum*. University of Chicago Press.

Dudley, S. (2021). *Displaced things in museums and beyond: Loss, liminality and hopeful encounters*. Routledge.

Ferrie, C., & Lewis, G. F. (2021). *Where did the universe come from? and other cosmic questions: Our universe, from the quantum to the cosmos*. Sourcebooks.

Foti, P. (2018). *Collecting and exhibiting computer-based technology: Expert curation at the museums of the Smithsonian Institution*. Routledge.

Grumbling, E., & Horowitz, M. (2019). *Quantum computing: Progress and prospects*. The National Academies Press.

Jaeger, L. (2018). *The second quantum revolution. From entanglement to quantum computing and other super-technologies*. Springer.

Jochim, B. (2022). *From NFTS to quantum computing*. www.rightclicksave.com/article/from-nfts-to-quantum-computing

Tatnall, A., Blyth, T., & Johnson, R. (2013). *Making the history of computing relevant: IFIP WG 9.7 international conference, HC 2013, London, UK.* Springer.

Weber, M. (2016). Self-fulfilling history: How narrative shapes preservation of the online world. *Information & Culture, 51*(1), 54–80.

Yale Quantum Institute. (2022). *'The quantum revolution' lands at the New Haven Museum.* https://quantuminstitute.yale.edu/publications/quantum-revolution-lands-new-haven-museum

Provocation no. 3: why is the computer different?

Kimon Keramidas

We notice that museums are still processing the dilemma of actually determining what, say, a 'personal computer' is. After all, a computer is not a computer until it is on and being used. A personal computer is a tool meant for cybernetic engagement with a human being; the human inputting, the computer responding, providing feedback. The museum's challenge therefore is not just to collect the object but to capture this reciprocal and inherent human action as well. What is problematic here is that not only are these actions and experiences ephemeral, but the personal computer is constantly changing in response to them. Every update changes the software, which changes the interactivity with (and experience of) the user. Every new security patch, every new widget, every new emoji – each alters the object and the everyday experience of that object. All of these assets aggregating into a dynamic system.

Take your smartphone for example. Your experience of that device is fundamentally unique as it is determined by the apps you install on the device and your personal interface preferences. Customization defines the object. The device is engineered to enhance the participation of the individual within frameworks determined by the persistent spending economy of late capitalism. Furthermore, as late capitalism expands more fully into an era of experience and attention economies, the device becomes a lifestyle choice for each user, and for many deeply embedded in one's sense of self. How does a museum, therefore, collect that framework, that dynamism, those ephemera, those choices? To collect a personal computer is to confront the collecting of mass marketing, mass sales, a mass of users. It is to collect a device that is much more deeply engrained in far reaching networks of lived experience than many of the objects normally collected and displayed by museums. And so how does the museum collect the devices created by a phenomenally corporate driven sector, by some of the wealthiest corporations on the planet right now (objects that should never not be part of the history of mankind at this point), but at the same time do so in a way that is always self-aware – noticing and explaining how so purposely self-obfuscating the late-capitalist framework within these objects is.

DOI: 10.4324/9781003424703-13

Alongside the dilemma of properly collecting the market and user experiences that surround personal computers, museums are further confronted with the conundrum of how to collect the expansive networks of data that flows in and out of these devices – and that in doing so define their use. In other words, how would a museum collect an Oracle network, a Cisco router system, a healthcare database, or an online social group – given that all of these are a profound part of contemporary life for many people today? What single element of each of these systems would be isolated as a collectable 'object'? And perhaps more pressingly, given in reality, where such data sits, and who owns such networks, would any of this be collectable anyway?

The answer to this dilemma – or at least the beginnings of an answer – will rely upon museums' willingness to break from usual practices. In its user operation, its societal pervasiveness, its market influences, the computer is just different to so much (if not all) of what museums confront and collect in the world. And so, when engaging with computers as heritage, museums need the courage to break from orthodoxy. Computer history demands new ways to deal with a set of objects which, for the museum, are unusual in its scope, scale, and complexity.

The information age has brought with it one of the most profound moments of societal change. Compressed into the personal computer is the enormity of this change – as well as the enormity of a new curatorial challenge. What is clear is that our conversations and actions around the collecting of computer history should never stop at the border of these devices.

Part IV

Lived practice of computing history

7 The CHM stack

Experimentation for digital and computing heritage

David C. Brock, Hansen Hsu, Dag Spicer, and Marc Weber

At the Computer History Museum (CHM), both our digital and physical collections are foundational. They are used to both build and to challenge diverse narratives. UNESCO describes 'digital heritage' as "computer-based materials of enduring value that should be kept for future generations" (UNESCO, 2009). In addition to digital heritage – which includes both software and digital objects of many types – CHM collects a more straightforwardly physical heritage of computing: three-dimensional objects, audiovisual materials, manuscript archives, rare publications, and oral histories. CHM's digital heritage collection has presented many new challenges and opportunities. In response, we have undertaken a broad range of activities, many of them experimental.

Today's digital objects have little meaning in isolation, without the computing environment and knowledgeable users needed to make sense of them. As they age into digital heritage materials, reproducing the socio-technical ecosystem in which they are embedded becomes ever more complex, implicating the people and practices necessary for migration, restoration, reconstruction, documentation, and long-term preservation. Attending to this ecosystem for digital heritage also makes clearer the kind of socio-technical requirements that *physical* heritage demands, as the recent work of Fernando Dominguez Rubio has so masterfully revealed (2020, 2018).

Computing communities often view systems hierarchically, as a kind of technical 'layer cake.' A so-called software 'stack' extends upwards from the operating system layer at the bottom to the applications layer that users interact with at the top. Each layer depends on the functions of the layer beneath. The networking layer model provides another pertinent example. This chapter reviews the development of a 'CHM stack' for contending with computing heritage. It details how levels of this stack operate on different timescales, from the permanent collection to ephemeral events, and how activities in collection, preservation, and interpretation, fundamentally rely on various communities and craft practices. The chapter concludes by describing OpenCHM – a literal rather than metaphorical software stack – being built by CHM and its partners to organize data in and about CHM's content and to provide more open access.

DOI: 10.4324/9781003424703-15

Time frames and the CHM stack

Any collecting museum can be seen as a stack of activities operating at different timescales, from the permanence of the collection to ephemeral events and social media posts. However, this layered structure can also play a role that may not be obvious at first – to allow for a diversity of narratives. Dominant interpretations within a field of expertise change over time. For the history of computing, these have shifted from technical and teleological narratives about a few favored systems and individuals, to a more inclusive mix of stories about both makers and users. Having said this, major gaps in understanding remain. Narratives are also influenced by the voices of pioneers of technology who are still alive today, and of course by popular coverage (Weber, 2015; Ensmenger, 2012; Mahoney, 1988; Novick, 1988). These changing stories directly affect what source material gets valued – and thus preserved. This issue has been foregrounded by the ephemerality of digital media, which is so central to the current age. Documents are still being discovered that have laid dormant in attics for decades, which may support new interpretations of the past. This is far less likely for disks or tapes and seemingly impossible for cloud-based storage. Early biases can become literally self-fulfilling, by erasing the source materials that could support competing points of view.

Partly by design, CHM adheres to no single historical narrative or frame but rather a series of working narratives for different functions of the institution. Much of the variety of these narratives has to do with the time frame, or the 'shelf life' of the material handled by those different museum functions. Events and blogs can last weeks, exhibits months to years, permanent collections ideally for centuries. Because events and exhibits generate materials for the permanent collection, they act as a hedge against biases in formal collecting.

At CHM, public events and blogs are ephemeral, although many enjoy a long tail of usefulness. They can be as topical as journalism as long as they have some historic grounding. Events and blogs can reflect nearly any type of interpretive framing, from the social history of gaming, to the biography of tech notables. Recordings of events can become part of the permanent collection; moreover, events can stimulate interest and lead to opportunities to collect other related materials, from oral histories to objects and archives.

Exhibits have a longer shelf life, ranging from decades to a few months. Many add valuable materials to the permanent collection as part of their creation. Exhibits need to entertain, of course, but this helps to encourage the coverage of a variety of narratives. Panels focused on the biography of individuals alternate with panels focused on social context, or portraits of systems and users from different parts of the world. Some exhibits focus on the story of the makers, such as our permanent exhibition *Revolution: The First Two Thousand Years of Computing*; others focus on how users shape technology, such as the exhibition *Make Software, Change the World!*, which includes stories about the social roles of texting, World of Warcraft, Wikipedia, and MP3, amongst others. While in development, all exhibits receive input from a mix of in-house stakeholders and outside subject matter experts. This variety helps ensure a richer range of narratives.

How much does the history of failures matter? Who are more important, inventors or implementers? Are entrepreneurs merely puppets for larger forces, or are they as pivotal as they like to think? For a collecting museum these are not just idle philosophical musings. Along with other academic questions, they feed fundamentally into decisions about collecting policies, budgets, and staff resources. Even the best-thought-out exhibit is a product of its time. But the permanent collection is intended to be the longest-lived level of museum activities. It is a resource sent forward to tomorrow's researchers, journalists, curators, and visitors. This means that the collection needs to be able to support a variety of interpretive frameworks, both current and future. CHM has formal sessions to define the scope of the collection, which bring questions about interpretive frames into sharp relief and can help fill in gaps. As an illustration of this, in recent years CHM has increased its intake of oral histories and other material from BIPOC computing pioneers.

How CHM built its stack

CHM began life in 1979 with a founding artefact, the legendary 1951 Whirlwind computer from the Massachusetts Institute of Technology (MIT). Digital Equipment Corporation (DEC) co-founder Ken Olsen and vice president of engineering Gordon Bell had both graduated from MIT in the mid-1950s and knew Whirlwind well, as it had profoundly influenced their work, as well as computing more broadly. In the late 1970s, Whirlwind was finally decommissioned by a private individual working for the navy who owned the machine. When Olsen and Bell heard that Whirlwind was destined for disposal, they took advantage of their financial and organizational resources and "turned the truck around," arranging for Whirlwind to be brought to DEC's corporate campus in Massachusetts (Bell, 2011; The Computer Museum, 1980–1998).

With this Ur-object acting as an impetus, Olsen, Bell, and colleagues began collecting some of the most important objects in computing history, including several of the first generation of electronic digital computers. They were not solely collecting hardware but also all known software associated with Whirlwind – the museum's first digital heritage materials – which existed on paper and magnetic tapes, from which CHM was able to recover digital information decades later. From the early DEC campus days, the collection became, again with DEC support, the central asset of a nonprofit public museum. This incarnation was named the Computer Museum, located on Boston's popular Museum Wharf, and attracting approximately 50,000 visitors per year (Computer History Museum, 2014).

In 1995 the collection was moved to Silicon Valley, California, providing the basis for a new nonprofit public museum, the Computer History Museum. The museum was intended to help capture the history of this unique region and to inform and inspire the world-shaping developments produced by its local tech communities. It soon became clear that along with the vastly increased stream of donations unlocked by the new Silicon Valley location, vital historical materials were both available and at risk. In response, the museum invested in saving these critical contextual documents, and eventually this sub-collection grew into a fully established archive, with nearly a linear mile of documentation, kept in a separate building and administered by a staff of five full-time archivists. CHM also established a small reference library – so it is engaged in all four of the unifying GLAM (Galleries, Libraries, Archives, Museums) concepts (Computer History Museum, 1999–2019).

Once in Silicon Valley, the pace of software collection increased dramatically. Several special interest groups (SIGs) devoted to software and its preservation were formed. These groups boasted impressive memberships, including professionals such as senior computer programmers, professors of computer science, software engineers, historians of computing. These efforts resulted in important new directions for software preservation at CHM, with "preservation" meaning different things in each group (Grad & Johnson, 2012).

For certain studies of software, the most useful form was generally agreed to be its source code – this is the part of software that people actually write and read and which can reveal the conceptual, functional, and social organization of the code. However, software in its executable form was also understood as vital for restorations of historical computer systems and demonstrations of their functions and capabilities. An alternate means of preservation that has received less attention than archiving code is the recording of user experience. By observing the interaction between user and their software, historians of computing have acquired valuable insights into how software functions both technologically and culturally. CHM has undertaken groundbreaking work in this area and considers the video recordings that document user experience to be a vital part of institutional software preservation.

However, perhaps the most important step made by CHM in addressing the history of computer software was the creation of the Robert M. Miner Curatorship in Software History, an endowed chair made possible by the sisters of the late Oracle co-founder Robert M. Miner. CHM curator Al Kossow, with four decades of software and hardware experience, including 30 years as an engineer at Apple, became the inaugural software curator, a position he holds to this day. With this position, the museum increased its software collecting activities. Kossow created an impressive online archive of computer software and computer documentation at bitsavers.org. The site contains millions of pages of documentation and extensive software holdings and has become a permanent part of the US Library of Congress Web Archive (Bitsavers' Software Archive, 2010).

During the mid-2000s the Web History Center, which was partly based on the original Web History Project of 1996, began working with CHM. This led to the founding of the CHM Internet History Program in late 2008. The program brought in expertise to address the challenges of collecting what we now understand as 'cloud' software. Various types of online and mobile software, including the code in the CompuServe archives, were also collected (Weber, 2020; Ahlstrand, 2022). In 2012 CHM began developing a major software exhibition, *Make Software, Change the World!*. This focused on the social role of software and the stories of users as well as makers. The exhibition's seven galleries included a mixture of stories, from Photoshop to World of Warcraft to SMS texting, with an eighth educational gallery for explorations of the nature of software (Computer History Museum, 2019a). The exhibition opened in 2017.

The CHM stack in action

CHM now stewards the world's leading collection centered on computer history, chronicling the remarkable impact of its innovators and inventions on human history. Our comprehensive collection of artefacts, documents, photographs, moving images, digital objects, and oral histories, are central to the museum's activities, including our permanent and temporary exhibitions, as well as our educational and programming activities. The collection covers a vast array of iconic computers and companies, from an Apple 1 to the latest trillion-transistor AI chip. Our invaluable set of more than 1,100 oral histories brings personal stories to light (Computer History Museum, 2020).

In the normal operation of the museum, layers of the stack interact with others in multiple ways. Higher, shorter-term layers such as blogs or events frequently enrich and promote lower, longer-term layers like exhibits, or reveal and interpret the lowest layer of the stack, the permanent collection. In turn, connections that are strengthened through exhibits, events, and blogs lead to new opportunities and avenues for collection.

By 2016, CHM had strengthened activities and resources at every layer of the stack to help deal with the unique challenges and opportunities of collecting and displaying digital heritage. Topic-based, long-term activities like the Internet History Program and the Software History Center, discussed next, cut across every layer of the CHM stack. The following case studies explore the stack in action around digital heritage and the history of software.

A software history center

In 2016 CHM established its Software History Center and hired two additional curators. Its purpose was to expand the collection and interpretation of software as a complement to the major exhibition *Make Software, Change the World!*, which had been in development for over four years. Both the exhibition and the Center signaled the increasing importance of software in the lives of people around the world and acted as correctives to a largely outdated perception of CHM as a museum primarily focused on hardware. Over the past seven years, the curators at the Software History Center have performed tasks parallel to those of their partner curators responsible for other domains – collecting historical materials including software and documents, conducting research, organizing events, and publishing in peer-reviewed journals. Additionally, the Software History Center was charged with tackling the challenges of digital heritage preservation and interpretation in the museum context and to consider new approaches that could serve CHM's diverse range of constituencies: tech fans, history buffs, students, tourists, scholars, museum lovers, and families. Eventually, the Center recognized that combining two areas of deep existing expertise at CHM – oral history and restorations – to produce video ethnographies of software would offer a powerful synergy.

Oral histories

CHM keeps an extensive and active oral history collection, with over 1,100 interviewees representing a broad spectrum of contributions to the story of computing. Most of these interviews were recorded on video. When completed, CHM makes both the video and the transcripts available to the public online. These videos have proven invaluable historical resources both for the work of the museum and externally for researchers, scholars, writers, and content producers of all types.

Today CHM's oral history program, alongside the work of other museums covering the history of computing, faces a generational challenge. Participants in computing who began professionally in the 1960s and 1970s are now entering their 70s and 80s. Many are now disappearing, although this generation of computing contributors is still vastly larger than the founding generation of the 1940s and 1950s. Without a wide-scale expansion of oral history activity, an incalculable wealth of historical information and insight will be lost forever. The generation who entered computing in the 1980s and 1990s is larger still. This further reinforces the need for expanding our capacity for recording oral history. At CHM, we are taking early steps in this direction by exploring how we can develop our program into a platform for other researchers pursuing oral history and through partnership with organizations such as the Association for Computing Machinery.

Restorations

Some museums devoted to the history of computing are focused on exhibiting working, often extensively restored, computing devices. Notable examples include the National Museum of Computing in the UK, the VCF Museum at the InfoAge Science Center in New Jersey, and the currently closed Living Computer Museum in Seattle. At CHM, we also engage in restoration of computers in our collection, but on a case-by-case basis and to serve specific goals.

For our onsite visitors, CHM has benefited enormously from a group of highly skilled volunteers – a great many of them computing pioneers themselves – who have carefully restored, continually maintain, and regularly demonstrate two iconic computer systems to the public: the IBM 1401, an extremely successful mainframe for business users, and the DEC PDP-1, a breakthrough minicomputer key to several milestones in the development of computer technology. For the research community, CHM has also benefitted from having the software curator Al Kossow as a member of staff devoted to software preservation, a practice that often necessitates the restoration of computer systems and peripheral devices. These restoration practices allow us to preserve historical software and other data and to make it available to researchers, hobbyists, and the public.

Video ethnography

The influential historian of computing Michael Mahoney stresses that the primary source for the history of software is a 'dynamic artefact,' that is to say, software running on a computer (Mahoney, 2008). Mahoney's observation has had a considerable impact on the Software History Center and CHM's strategy for the collection and interpretation of the story of software. Recent scholarship in the history of technology, science and technology studies (STS), and the history of computing has emphasized the critical role played by human actors in the creation, operation, and maintenance of technological systems (Bijker et al., 1987; Biagioli, 1999; Haigh & Ceruzzi, 2021; Mullaney et al., 2021; Abbate & Dick, 2022). The Software History Center has elaborated on Mahoney's conception of the dynamic artefact of the history of software, foregrounding a third element implicit in his formulation. For us, a dynamic artefact is constituted by properly functioning software, running on a fully working computer that is operated by a knowledgeable user.

While the notion of a dynamic artefact captures the idea of multiple components operating and interacting in real time, it is a concept that requires further elaboration. What makes this kind of artefact dynamic? For the Software History Center, the answer is *performance*. Inspired by anthropologists and others who have used moving images to document cultural performances, including users in virtual worlds, and adding to previous CHM work in recording inventors in the act of demonstrating web software and mobile and UI devices (Weber, 2013), the Software History Center has undertaken a series of projects in what we call *video ethnography* of the history of software (Shrum & Scott, 2017; Visual Ethnography, 2012–2022; Lowood, 2011; Lowood et al.,

2017). In this practice we combine oral history and computer restoration to video record a variety of demonstrations and explorations of historical software running on restored hardware, operated by knowledgeable users who are primarily the creators of the software. To date, we have conducted video ethnographies on a variety of programs and environments on the Xerox Alto, other software from Xerox PARC, and multimedia systems from the 1980s and 1990s (Computer History Museum, 2017a, 2017b, 2018a, 2018b, 2018c, 2019b). It is our ambition for these video ethnographies to serve as long-term research resources and as material that can be used for CHM exhibits, blogs, events, and other products serving its diverse publics, providing an opportunity to see and hear historical software in action.

The end of software collecting?

Collecting commercial, proprietary software that has been created in the contemporary era is becoming practically impossible. This is largely due to a shift in the industry towards software as a service via the cloud, via subscription models, or through proprietary app stores. Marc Weber's *Core* magazine article, "The Net is Eating Software" (2017), summarizes this new digital 'dark age' very effectively. Physical boxes and discrete version numbers have given way to continuous integration and deployment. Moreover, much software today is predicated on a connection to, and the availability of, network services. Indeed, some apps on your device depend upon a connection to someone else's working computer.

Our recent experience collecting the original Siri app provides a good example. Before its acquisition by Apple, Siri was a spinoff from SRI, a storied contract research firm, whose mobile app was downloadable on the iPhone App Store in 2010. The app was a voice-centered digital assistant that responded to the user's spoken queries through speech recognition, speech synthesis, and advanced search. CHM recently acquired a copy of the Siri app, recovering it from a period iPhone and copying it onto a contemporary system for the collection. However, there is no way to successfully run this software on a contemporary computer or smartphone. Even on its original iPhone, the Siri app fails to function. This is because the servers on which it depends for almost every feature are gone. Apple acquired the Siri company soon after its 2010 launch and retired its servers. Without them, the app will forever remain a functionless user interface.

Considering these difficulties, CHM approaches today's digital heritage with the holistic strategy of collecting everything relating to the software and the people who created or used it, including documents, manuals, audiovisual materials, oral histories, and video ethnographies. Today, CHM is working to develop new levels of access to these materials, and to our collection more broadly, through the OpenCHM initiative.

OpenCHM

OpenCHM is a strategic, long-term initiative designed to build a technology platform that can unlock CHM's collection and make it broadly accessible across the globe. It is intended to provide an entirely new degree of access to CHM's oral histories, archival collections, videos, photographs, historic software, and past events on a rich variety of topics. OpenCHM will also offer transparent access to digital assets that are *not* part of the permanent collection, such as content brought in for blogs, events, or exhibits. This is just one way it will make interactions between layers of the CHM stack even easier.

CHM staff are currently working with corporate partners Terentia and Microsoft, with significant support from the Gordon and Betty Moore Foundation, to implement a new backend collections management system that will underpin access to the new online collection platform. This platform will allow CHM staff to manage all aspects of their collection, exhibits, and digital assets and will also provide visitors with new ways to access the collection online. OpenCHM will enhance browse and discovery features through the integration of state-of-the-art search functionality. Online browsing options will be enhanced by curated sets of material centered around a given theme, and digital stories that connect directly back to the collection. Moreover, CHM is embarking on a mass digitization project to provide online access to a larger portion of the collection (Brock, 2022; Spicer, 2022).

Conclusion

CHM's long engagement with the preservation and interpretation of physical and now the digital and archival heritage of computing has led the organization to engage in a broad array of interconnected activities, operating at different timescales and intertwined with diverse narratives. Continuing developments in the nature and use of computing in the contemporary moment and into the future will doubtless offer new affordances for preservation and interpretation, just as they will create new dilemmas. We hope the continued evolution of the CHM stack and the relationships between its layers can provide a buffer against future biases, create new tools to address fresh challenges, and ensure a variety of inputs and points of view. This in turn can help ensure that our long-term contribution to society is as much about people, communities, and narratives, as about devices, code, and systems.

Reference List

Abbate, J., & Dick, S. (2022). *Abstractions and embodiments: New histories of computing and society*. Johns Hopkins University Press.

Ahlstrand, P. (2022). *Mission impossible, CHM edition: Rescuing the CompuServe collection.* https://computerhistory.org/blog/mission-impossible-chm-edition/

Bell, G. (2011). *Out of a closet: The early years of the computer museum.* https://tcm.computerhistory.org/outoftheclosetV2.3.pdf

Biagioli, M. (1999). *The science studies reader.* Routledge.

Bijker, W., Hughes, T., & Pinch, T. (1987). *The social construction of technological systems: New directions in the sociology and history of technology.* MIT Press.

Bitsavers' Software Archive. (2010). *United States* [Web archive]. The Library of Congress. www.loc.gov/item/lcwaN0004602/

Brock, D. (2022). *A museum's experience with AI.* https://computerhistory.org/blog/a-museums-experience-with-ai/

Computer History Museum. (1999–2019). *Core: The annual magazine of the computer history museum.* https://computerhistory.org/core-magazine/

Computer History Museum. (2014). *The computer museum digital archive.* https://tcm.computerhistory.org/

Computer History Museum. (2017a). *Alto system project: Larry Tesler demonstration of Gypsy* [Moving image]. www.computerhistory.org/collections/catalog/102738551

Computer History Museum. (2017b). *Video ethnography of 'ICARUS' on the xerox alto.* www.computerhistory.org/collections/catalog/102738686

Computer History Museum. (2018a). *Alto system project: Dan Ingalls demonstrates Smalltalk* [Moving image]. www.computerhistory.org/collections/catalog/102738723

Computer History Museum. (2018b). *Video ethnography of visual almanac* [Moving image]. www.computerhistory.org/collections/catalog/102738853

Computer History Museum. (2018c). News navigator video ethnography [Moving image]. www.computerhistory.org/collections/catalog/102738712

Computer History Museum. (2019a). *Make software: Change the world!.* https://computerhistory.org/exhibits/make-software/

Computer History Museum. (2019b). *Eric Bier demonstrates cedar.* www.computerhistory.org/collections/catalog/102781041

Computer History Museum. (2020). *Oral histories.* https://computerhistory.org/oral-histories/

The Computer Museum. (1980–1998). *Annual reports.* https://tcm.computerhistory.org/reports/TCM_Annual_Reports_1980_1988-1998.pdf

Ensmenger, N. (2012). The digital construction of technology: Rethinking the history of computers in society. *Technology and Culture, 53*(4), 753–776. Johns Hopkins University Press.

Grad, B., & Johnson, L. (2012). Collecting the history of the software industry. *IEEE Annals of the History of Computing, 34*(4), 85–87. https://doi.org/10.1109/MAHC.2012.57

Haigh, T., & Ceruzzi, P. (2021). *A new history of modern computing.* MIT Press.

Lowood, H. (2011). Video capture: Machinima, documentation, and the history of virtual worlds. In H. Lowood & M. Nitsche (Eds.), *The machinima reader* (pp. 3–22). MIT Press.

Lowood, H., Kaltman, E., & Osborn, J. (2017). Screen capture and replay: Documenting gameplay as performance. In G. Giannacchi & J. Westerman (Eds.), *Histories of performance documentation: Museum, artistic, and scholarly practices* (pp. 149–164). Routledge.

Mahoney, M. (1988). The history of computing in the history of technology. *IEEE annals of the history of computing, 10*(2), 113–125. https://doi.org/10.1109/MAHC.1988.10011

Mahoney, M. (2008). What makes the history of software hard. *IEEE Annals of the History of Computing, 30*(3), 8–18. https://doi.org/10.1109/MAHC.2008.55

Mullaney, T., Peters, B., Hicks, M., & Philip, K. (2021). *Your computer is on fire*. MIT Press.

Novick, P. (1988). *That noble dream: The "objectivity question" and the American historical profession* (pp. 12–13). Cambridge University Press.

Rubio, F. D. (2018). On the discrepancy between objects and things: An ecological approach. *Journal of Material Culture, 21*(1), 59–85. https://doi.org/10.1177/1359183515624128

Rubio, F. D. (2020). *Still life: Ecologies of the modern imagination at the art museum*. University of Chicago Press.

Shrum, W., & Scott, G. (2017). *Video ethnography in practice*. SAGE Publications.

Spicer, D. (2022). Computer history museum, fall 2021 update. *IEEE Annals of the History of Computing, 43*(4), 108–110. https://doi.org/10.1109/MAHC.2021.3121591

UNESCO. (2009). *Charter on the preservation of digital heritage*. https://unesdoc.unesco.org/ark:/48223/pf0000179529.locale=en

Visual Ethnography. (2012–2022). www.vejournal.org/index.php/vejournal/index

Weber, M. (2013). Exhibiting the online world: A case study. In A. Tatnall, T. Blyth, & R. Johnson (Eds.), *Making the history of computing relevant: IFIP WG 9.7 international conference, HC 2013, London, UK. IFIP advances in information and communication technology* (Vol. 416, pp. 3–4). Springer. https://doi.org/10.1007/978-3-642-41650-7_1

Weber, M. (2016). Self-fulfilling history: How narrative shapes preservation of the online world. *Information & Culture: A Journal of History, 51*(1), 54–80. University of Texas Press. https://doi.org/10.1353/lac.2016.0003. (Original work published 2015)

Weber, M. (2017). The net is eating software. In *Core: The annual magazine of the computer history museum* (pp. 57–61). https://computerhistory.org/publications/core-magazine-2017/

Weber, M. (2020). *Internet history program*. https://computerhistory.org/internet-history-program/

8 Beyond *point and click*

Calling out expediency in museums' histories of computing

Lisa McGerty

The Centre for Computing History (CCH) began in 2006 when the idea for a new 'hands-on' museum of personal computing was conceived by three friends, none of whom were museum professionals but one of whom owned an extensive collection of computing-related artefacts. The idea developed, we sought consultation, formed a charity, raised funds, and opened the museum to the public in Cambridge, UK, in 2013. The museum is now accredited, enjoys an active visitor base, and engages in the presentation of public history like any typical museum. This first iteration of a computer museum in Cambridge, an obvious location for a museum of personal computing in the UK, was viewed as a 'beta test' by the workforce, although we now know the museum is feasible, both financially and conceptually. So far, so good. The next step, having proved that CCH is viable, is to add 'shine' so the museum can one day become a world-class visitor attraction with a well-documented collection. This has been our aim since the very beginning. In short, The Centre for Computing History is currently undergoing the process of 'growing up' in public.

But what sort of history is CCH engaged with? As a museum matures, so do its restraints – resources, business models, core audiences – and at various times, crucially, those restraints also generate opportunities – to be experimental, responsive, agile, open – and all of these things inevitably change over time. Fred Wilson suggests that "the bigger museums get the more they lose their nimbleness" (Marstine, 2013, p. 39), but CCH is still relatively young and 'nimble'; in other words, we still have the freedom of limited resources, though such 'freedom' is inevitably embedded in our own very particular values and ways of working.

Using three examples of exhibitions as a lens through which to explore CCH's values and museum practice, in this chapter I explore, as a feminist museum practitioner, how the 'actuality of the museum', that is to say, the organizational context within which CCH operates, impacts the histories of computing that we produce, present, and perform. I then suggest that this chapter might act as a step in our journey of growing up, by undertaking a critique of our own practice, a process through which we may be able to embrace

DOI: 10.4324/9781003424703-16

different values, notice different things about computing history, and thereby produce new histories in the future.

Convergence: interactivity and CCH as a nonprofit business

CCH's core objectives are to preserve our collection and use it to create learning experiences by telling the stories of the Information Age through exploring the historical, social, and cultural impact of developments in personal computing. When CCH was formally established in 2007, an entrepreneurial spirit seemed a prerequisite for getting a new museum of computing off the ground without significant public funding, and this spirit continues to exist in our workforce to some extent today, despite changes inevitably taking place as we mature as a museum. Although our workforce had little professional museum expertise, business and computing experience was not in short supply, and the business dimension to the origin story of CCH embodied some core values that continue to impact our museum practice today. These values have a profound influence on the histories of computing we produce and perform, in both a constraining and an enabling way.

One particularly salient aspect of this ongoing operational reality is that, although a charity, we have never been dependent upon financial support and prioritize being able to reliably pay our bills. That translates into museum practice that prizes visitor satisfaction, so as to keep funds flowing (if nothing else) and carrying out all the tasks of running the museum ourselves, so as to make the very most of generated funds. This means our exhibitions are generally built using our own labor and are designed to ensure a high level of visitor engagement. On a superficial level, this self-funding ethos means we can tell whatever stories we want to tell with the collection we have, as long as our workforce has the right skills and time to develop the exhibits, visitor feedback remains good, and our income keeps growing along with visitor numbers.

At CCH, high levels of visitor satisfaction are understood as stemming from our 'authentic' portrayal of computing; in other words, high levels of interactivity in the museum generate reliable revenue. In practice, like other museums whose collections consist of relatively modern, mostly mass-produced objects, we encourage visitors to handle and use the exhibits extensively. From the outset we reasoned that computers can be experienced most meaningfully and enjoyably when they are used rather than merely seen. Therefore, almost every exhibition and event since the very beginning of CCH has been 'hands-on,' using mostly working machines. This commitment to interactivity is central to everything we do, and it remains the case despite the Covid-19 pandemic, when 'touch' in public spaces was very much problematized. Without ever really articulating it, through this offer of a highly interactive museum we have tried to break the 'black box conundrum' (Foti, 2018), and to harness the sensory experience of computing technologies as museum objects so as to facilitate visitor engagement and learning (Baldwin, 2019; Howes, 2014; Howes et al., 2018). Put another way, we break down barriers between visitors and the objects on view (Baccaglini, 2018), on the premise that engaged visitors are happy visitors and happy visitors come back. The average amount of time a visitor spends at the museum is currently between three to four hours. Levels of income are healthy, and audience research indicates that the immersive, interactive experience we offer is valued highly by visitors.

In fact, so prominent is the principle of interactivity at CCH that interpretation was almost nonexistent when we first opened in 2013, and often remains minimal. We rely heavily on visitors to bring their personal experiences with them to help tell stories vis-a-vis the objects on display. This often operates as nostalgia for older visitors and as an opportunity to form new memories to look back on for both older and younger visitors. We stage encounters between old and new, as most museums do, and we do so by encouraging visitors to experience key stories or moments in the history of computing by handling objects, by taking in the accompanying interpretive material, and making use of the museum's visitor assistants. The idea that CCH is a museum of computing rather than computers is inextricably linked to this; in order to explore the history of computing, we are in essence exploring the human experiences with which the machines are entangled.

One exhibition, which we named *Convergence* and focused on the development of the smartphone, provides a particularly good example of this in action. Interpretation is minimal. The exclamations of visitors – 'I had one of those!'; 'What is that?!' – as they use old and new devices, provide the narratives and a substitute for written interpretation. Our decision to prioritize active visitor engagement over more traditional, 'didactic' forms of museum communication and learning is clear. Visitors explore the story of various technologies converging over time to eventually become the computer in your pocket, a.k.a. the smartphone. The exhibition offers a presentation of technological advancement in terms of the progression of computational functionalities through history, driven by commercial interests, in which new technologies essentially incorporate old ones. The narratives on display in the *Convergence* exhibition, then, imply, or even state explicitly, that new heights of human achievement are reached using new computer technologies.

Given that we have a relatively small team of staff to work on exhibitions, limited space, tight budgets, and an ethos that encourages visitors to interact with our objects, *Convergence* was both expedient and effective, as visitors could interact with various objects and in the process marvel at how we as humans used to live and how much progress we've made as a result of the technologies we've created. Displays such as these, which minimize more traditional forms of museum interpretation and the communication of the knowledge of our own experts, are possible because a certain level of visitor expertise is nonetheless achieved. Clearly though, the particular organizational and operational structures of CCH, give rise to such a heavily experiential visitor experience and impact the computing histories we are producing. In other words, the computing histories that are performed at CCH are shaped by a museum space in which visitors handle and experience artefacts and enjoy and make sense of them in their own way. We know visitors enjoy the immersive, interactive experience we offer, so we keep that as our core aim, and the type of history we represent inevitably reflects that too.

We are aware of the types of historical commentary that the *Convergence* exhibition could bring about, that is to say, the narratives about how convergence happened or the changes in our workplaces, homes, and families that made the smartphone a possibility. With *Convergence*, CCH doesn't tell those stories perhaps because we are particularly reliant on the visitors' participation, which emerges through the enjoyment of an 'authentic' computing experience. Our space has to be used well, not just because it is small, but because we must maximize enjoyment so that visitors are happy, spend while they're with us, and want to come back.

However, is there space within this approach for less tangible stories, such as how the humble telephone was uncoupled from specific places and instead carried by individuals? Or to explore how human communication has reached a place in which we now communicate many times a day using only images or 140 characters and a hashtag? These are difficult stories to tell through the type of sensory museology (Howes, 2014) aspired to by CCH. However, perhaps there is a risk in telling more difficult and abstract stories because we cannot afford to challenge our visitor base in case we lose those visitors and our museum becomes unsustainable. Just like larger museums, which may have very different although undoubtedly equally effective operational strategies, we are compelled to write a popular history, to tell an inviting story. Evidencing and experiencing human 'progress' by helping visitors to experience old technology while marveling at new technology is perhaps the most compelling (hi)story of all.

Stories of rapid technological change – defined mostly by the addition of functionalities concurrent with making technology easier to use and narratives of human progress – abound at CCH. We celebrate how much computers are changing, and how quickly, but we are also careful about how we frame their impact, for reasons that have nothing to do with computing history and everything to do with wanting to keep our visitors happy. In other words, we minimize the discussion of the potentially complicated impact of computers on human social experience, to avoid making visitors feel uncomfortable; instead we try, mostly through touch, sound and play, to create a sense of wonder at the size and power of both old and modern computers, of valves and silicon wafers, and of the megaprocessor in our foyer. As of yet, we do little to reflect on the possible costs of these technologies in terms of people, social relations, the environment.

Using Simone Natale's (2016) notion of the biographies of media we can see that by inserting new technologies into existing narratives, we offer familiar trajectories for museum visitors as a strategy through which they can preserve their sense of everyday life in the face of potentially destabilizing technological change (Striphas, 2009; Ballatore & Natale, 2016). Our visitors are "survivors of the past" (Kavanagh, 2005), the past being a time when computers either didn't exist or were so unwieldy that they needed a great deal of expertise to 'tame' and use. Since then computers have become smaller and more ubiquitous. Just look at how much humankind has been able to achieve with them. If we told different stories, would visitors come back again? And if not, would our young museum still be able to pay its bills?

Who are we?: personal computing a.k.a. point and click

Some of CCH's other exhibitions reflect a different kind of operational reality and expediency of running the museum. For example, we have been developing a new exhibition tentatively titled *Point & Click*, which will tell stories about the ways in which humans interact with computers. Early iterations of the display convey a chronology of events in which humans cajole computers to do things, first by programming using binary and machine code (the preserve of 'experts'), then via the mouse and graphical user interface (less expertise required), until we can swipe, pinch and tap, and speak to Alexa and Siri, requiring almost no expertise at all, as long as users are able to afford the devices, are able-bodied, and do not have an audible accent).

Entirely in keeping with our focus on the way computers have changed over a short space of time, and how much humans can achieve as a result, the draft display includes a written and diagrammatic timeline of notable historical moments, from Douglas Engelbart's 'Mother of All Demos,' to the arrival of wearable technology. The various stories on the panels are embodied in and represented by the touchable objects on display, ranging from PERQ, IBM, and Amstrad office machines, to Amazon Echo. The implication is that technological advancement and the use of computers makes everyday life easier, regardless of the expertise of the user. The first paragraph of the introductory text on the draft panel summarizes this: "The GUI ('gooey') makes using computers easy and modern ones are so intuitive that users need know almost nothing about computing to use them."

However, unlike *Convergence*, *Point and Click* cannot foreground the museum visitor and their experiences, precisely because the user interface (UI) is (or should be) almost invisible to users; the stories told here make the distributive expertise that is central to the *Convergence* exhibition display impossible to replicate for this area of the museum. While tech-savvy visitors may be able to reflect on different interfaces they have used over time, other visitors would likely not. So to appeal to everyone, the panel resorts instead to common narratives about how users benefit from commercially driven technological advancements that make human lives easier. The underlying story is one of a transition from large computers, used by teams of experts, to smaller machines that fit onto the desk of an individual without specialist computer expertise, and beyond that, to devices accessed via voices and fingerprints. Along the way, so the story goes, individuals gained access to conveniences such as ATMs, satellite navigation, and the internet, and these make our lives easier. "Imagine not having ATMs," the panel asks of visitors.

So the drafted design plans for the *Point & Click* display tell a story similar to *Convergence* – a 'shift' over time that represents a 'major leap forward' for human progress. The objects and interpretation signify that this shift began with Engelbart's ideas and the development of the Alto at Xerox PARC, which the panel says launched the 'personal computing revolution.' The exhibition thus draws on, and reinforces, popular narratives about how computers became personal. These narratives are particularly important to CCH as the story of how the computer became personal, and in this way ubiquitous, is CCH's raison d'etre and is what distinguishes us from other computer museums in the UK. Graphical user interfaces are central to the story of how computers became personal, and this is signified on a draft for one of the panels by an enormous arrow pointer image, which will draw visitors towards this central area of the museum, that is to say, towards the start of the story we tell.

Our focus on interactivity and resisting the 'black box' conundrum by keeping objects working and even by exposing what is inside them, and our positioning of those as a gateway to authenticity and visitor satisfaction, makes telling stories around user interfaces, which are 'invisible' phenomena within objects, particularly difficult. But this exhibition may also have been incredibly difficult for us to develop for another reason. As a museum of personal computing, the importance of user interface development to the progression of computing from specialist activity to the personal and the everyday is an essential story. In developing the exhibition we needed to 'freeze' time and commit to a story that we feel would engage visitors and satisfy our desire to tell stories that are useful and interesting. Inevitably in a self-funded museum, stories that have commercial appeal cannot be dismissed, but so critical is the story of how computing became personal to our identity as a museum that before we can make this exhibition work we likely also need to clearly establish what we want to say. As such a young museum – at time of writing we have not yet been open for a decade – our story has not yet been frozen in time any more than the story of the UI has. CCH exists to explore personal computing and user interfaces are key to that, so perhaps we have struggled with this display because telling the UI story is also telling a story about CCH? Factor in our focus on interactivity for visitors as a way of resisting the black box conundrum (and monetizing it) and it is no wonder we have been struggling to develop this exhibit! Despite our best efforts to ensure we display a broad and representative collection of objects, our need to tell a headline story about personal computing that differentiates us from other museums and attractions, played out through historical objects, inevitably shapes the histories of computing we exhibit.

The lure of the virtual in curating computing history

Looking to the present and the future, like many museums CCH is delving into the digital and the virtual, not just via digital collections and online audience engagement, but in the development of a virtual computer as part of a lottery-funded project. We are rendering LEO, a first generation electronic computer from 1950s Britain – the first computer dedicated to business functions – in three-dimensional form for interaction at the museum, as well as via a tablet device that can be used by the public at home. As there are no longer LEO computers in existence anywhere, we are 'rebuilding' one virtually, so visitors and non-visitors alike can experience the machine by 'walking around' it whilst at the same time benefiting from access to our LEO archive, in order to learn about early computers, and the quirky development of LEOs by a company – J. Lyons & Co. – known for making tea and cake and opening cafes in a postwar London still blighted by bombed-out buildings and food rationing (Centre for Computing History, 2023). This kind of activity takes 'hands-on' to a whole new level. As Fiona Cameron says, "the value of the 'real' increases when digitized, enhancing its social, historical, and aesthetic importance, owing to the resources required in the compilation of a three-dimensional rendering, and through distribution" (2007, p. 51).

The digital revolution in museums, in which the goal of harnessing augmented and virtual reality to benefit visitors and non-visitors is the new frontier, promises great things for museums in terms of widening access, widening participation, and telling new stories. For us it also offers new ways for visitors to experience and enjoy 'old' computers, like the 'extinct' LEO I of 1950s London and to enable them to access an archive in a new way. Such work hints at possible new futures in which museums produce and perform histories of computing in new ways.

Reflecting on the funded projects of CCH also offers insights into how the realities and experiences of running any museum shape those histories. For CCH as a small, largely self-funded museum, project funding and working in partnership allows us, and allows me as a feminist practitioner, to carve out the space and time to produce new histories of computing in new ways, in this case by developing a digital reproduction of a computer that was particularly inventive and had an unusual social history. Such work may also be key in helping CCH to navigate our way towards a future in which our own internal operational environment and ideas of an 'authentic' experience of computing in the context of a museum may well develop into something completely different to what we currently do. Project funding therefore helps us overcome restrictions of time and resources, allowing us new opportunities for original exhibition-making. This also means we can afford to subvert our usual demands on visitors that require them to bring human stories that make sense of the computing artefacts they encounter at CCH and instead – by recognizing that few visitors would have experienced a computer like the LEO – to provide those stories in a way more akin to traditional museum storytelling, but wrapped up in new interactive technologies that give agency to visitors in different and nonlinear ways. This particular presentation of computing history is made possible by localized funding opportunities, coupled with the passions of individual curators. The funding allows us to transcend the usual restraints of resources, business models, and obligations to core audiences and to explore strategies that are more experimental, agile, and open, producing different histories of computing in the process. Thinking about our museum practice in this way shines a light on how our organizational and operational expediencies might be shaping histories of computing.

Perhaps museum objects such as the virtual LEO, that is to say, digital copies of the real thing, that explore an object's social history in an experiential way, will be the way computing history can offer vitality for audiences in the future? Where will the interactive museum that is CCH be once we have grown up and none of our visitors can remember brick phones or desktop computers – let alone mainframes or glass valves – and so cannot provide their own narratives to understand the displays of technological fossils we provide? Or what will happen when computers are so ubiquitous that we can't tell where the human and the machine each begin and end? Presumably we will need to tell different stories in different ways.

So what will CCH's future look like in terms of visitor experience? What if we can no longer keep the computers we display working? How likely is it that we can make a 60-year-old first-generation iPad function and survive handling, as we currently can with the 1960s Elliott 903 that we still run for demonstrations? Should we assume that we won't need to because the museum visit experience will be something entirely different by then anyway? Will it only be 'virtual'? Will all our museum objects need to be digital copies of the real thing, with the real thing only existing in a perspex case much like dinosaur bones do now? In which case, will they still make sense? Are we at risk of telling only part of the story of the *Information Age*, if it is only through the parts that we can keep working? Asking such questions helps us notice the multifarious ways in which the nature of CCH, our particular history and ways of working, and indeed the actuality of any museum, shape the histories of computing produced and performed. The different facets of the reality and expediency of running a museum are numerous and range from business models, organizational histories, the prioritization of visitor experience, the museum's particular positioning in the 'market' of visitor attractions, ideas of being experimental and open by exploring new methods such as virtual and augmented reality as a new way of accessing collections and, in our case, our confidence in having a core audience that can interpret the technological objects we display with often minimal interpretation. By necessity, all of these things change over time, so just as there is no 'essential museum' that is produced in the same way at all times (Hooper-Greenhill, 1992), there is no essential Centre for Computing History; our identities, targets, functions, and subject positions are variable and possibly discontinuous. All small museums struggle to tell their stories due to practical restraints, although these restraints can also be seen as opportunities, and the history CCH engages with, and the ways in which our museum therefore attempts a 'version control' of the past (Hoyle, 2017), is inevitably subject to constant change as powers, individuals, opportunities, and risks wax and wane as we grow up in public. However, as Mar Hicks points out, noticing these things now and critiquing our own actualities and practice so that we are not "functionally asleep at the wheel" may help provide the way forward to producing new histories for new technologies in the future (Hicks, 2021, p. 12).

Acknowledgment

This chapter was written with support from Jennifer Bergevin, to whom I extend my deepest thanks.

Reference List

Baccaglini, A. (2018). *Multi-sensory museum experiences: Balancing objects' preservation and visitors' learning* [MA thesis, Seton Hall University].

Baldwin, L. (2019). The experience of handling museum objects. *Royal College of Music Museum Blog*. www.rcm.ac.uk/about/news/all/2019-01-03museumblogtheexperienceofhandlingmuseumobjects.aspx

Ballatore, A., & Natale, S. (2016). E-readers and the death of the book: Or, new media and the myth of the disappearing medium. *New Media & Society, 18*(10), 2379–2394.

Cameron, F. (2007). Beyond the cult of the replicant: Museums and historical digital objects – traditional concerns, new discourses. In F. Cameron & S. Kenderdine (Eds.), *Theorizing digital cultural heritage: A critical discourse* (pp. 49–75). MIT Press. https://doi.org/10.7551/mitpress/9780262033534.003.0004

Centre for Computing History. (2023). www.computinghistory.org.uk/sec/46234/LEO-Archive-at-CCH/

Foti, P. (2018). *Collecting and exhibiting computer-based technology: Expert curation at the museums of the Smithsonian Institution*. Routledge.

Hicks, M. (2021). When did the fire start? In T. S. Mullaney, B. Peters, M. Hicks, & K. Philip (Eds.), *Your computer is on fire* (pp. 11–27). MIT Press.

Hooper-Greenhill, E. (1992). *Museums and the shaping of knowledge.* Routledge.

Howes, D. (2014). Introduction to sensory museology. *The Senses and Society, 9*(3), 259–267.

Howes, D., Clarke, E., Macpherson, F., Best, B., & Cox, R. (2018). Sensing art and artifacts: Explorations in sensory museology. *The Senses and Society, 13*(3), 317–334.

Hoyle, V. (2017). Archives and public history. *Archives and Records, 38*(1), 1–4.

Kavanagh, G. (Ed.). (2005). *Making histories in museums.* Bloomsbury Publishing.

Marstine, J. (2013). Museologically speaking: An interview with Fred Wilson. In R. Sandell & E. Nightingale (Eds.), *Museums, equality and social justice* (pp. 62–68). Routledge.

Natale, S. (2016). Unveiling the biographies of media: On the role of narratives, anecdotes and storytelling in the construction of new media's histories. *Communication Theory, 26*(4), 431–449.

Striphas, T. G. (2009). *The late age of print: Everyday book culture from consumerism to control.* Columbia University Press.

Provocation no. 4: decolonizing computing histories in museums

Lara Ratnaraja

Historical objects are often imbued with a meaning beyond their time-bound contexts, providing us with multiple narratives that span social, technical, and personal dimensions. Computers have had a seismic impact on society, and as outputs of the industrial complex, they bring with them complex and inter-sectional contexts and impacts. This can be seen in the tools and instruments of agricultural development, the Industrial Revolution, and now during the information age. However, the unique differential here is the relational aspect of the technology to the organization or institution itself.

Computers have had a huge industrial impact with regards to their use in production and how we now work. However, individuals working within any organization, including museums and cultural heritage institutions, have a personal connection with computers and smartphones, and they use them as sources of content consumption as well as creation. We are about to see further impacts with the arrival of AI software such as ChatGPT.

These technological developments have of course impacted us on an environmental level. Moreover, the history of the development of technology encompasses the lived experiences of those who contributed to, or simply experienced – and for the first time this is reflected not only in terms of class and socio-economics, such was the case with the agricultural and industrial revolutions, but in the identity of people from marginalized and underrepresented backgrounds, in terms of ethnicity, gender identity, sexuality, and disability. As a consequence, the responses to exhibitions on computing in museums are augmented by the psychological and cultural impacts technology has on us and by extension the organization.

Often this is at odds with the wider societal context and interplay of history, with museums developing within and responding to the frameworks of colonialism and capitalism. The far-reaching impact of these have skewed and affected interpretations of value and resulted in bias in coding and application.

Computing itself is not binary, and the influence of a plurality of lived experiences is also at play within the organizational museum context. Within the institution, curators and archivists are not truly objective about the

DOI: 10.4324/9781003424703-17

instruments of technology that they have a personal relationship with, thus skewing an objective understanding of the artefact and its historical significance. As Ross Parry points out,

> It has been a time of the sector articulating and evidencing the effect that the arts and heritage can have on society, reflecting on the ways the outputs and provision of these organisations can be socially inclusive, and of demanding a social diversity in the sector's workforce.
>
> (Parry, 2018, p. 35)

Museums need to unpick the narratives they have hitherto presented through the lens of colonialism and capitalism, and do more to decolonize narratives of computing, and create inclusive, resonant, and relevant computing histories. Museums that engage with histories of digital media and computing, like all kinds of cultural heritage institutions, need "to open up and excavate our institutions, dig up our ongoing pasts, with all the archaeological tools that can be brought to hand, sometimes a teaspoon and toothbrush, other times a pick-axe or a jack-hammer" (Hicks, 2021, p. 19).

In doing so, museums must look to a new form of blended narratives that accommodate multiple understandings, experiences, biases, and interpretations. Shifts in attitudes towards race, gender, sexuality, and disability, as well as diversity and inclusion, are key to this. As postcolonial perspectives have unpicked taken-for-granted truisms regarding certain objects and opened up a more a relational aspect to objects from a previous imperial era, for example, lived experiences and their relationship to objects and their contexts necessarily impact culturally on institutions and hence on how they tell those stories. As Sathnam Sanghera notes,

> Imperialism is not something that can be erased with a few statues being torn down or a few institutions facing up to their dark pasts; it exists as a legacy in my very being and, more widely, explains nothing less than who we are as a nation.
>
> (2021, p. 1)

The history of computing technology presents an opportunity to understand that stasis is no longer the norm in the display of historical objects. In order to present authentic and contextual narratives, museums must now radically address how the interplay of history and society plays a significant role in the narrative of computing heritage in museums.

Reference List

Hicks, D. (2021). *The Brutish museums: The Benin bronzes, colonial violence and cultural restitution*. Pluto Press.

Parry, R. (2018). *Socially purposeful digital skills*. www.keepandshare.com/doc/8226734/let-s-get-real-6-culture-24-rgb-single-page-pdf-10-5-meg?da=y

Sanghera, S. (2021). Empireland: How imperialism has shaped modern Britain. Viking.

Index

For Product Safety Concerns and Information please contact our EU
representative GPSR@taylorandfrancis.com
Taylor & Francis Verlag GmbH, Kaufingerstraße 24, 80331 München, Germany

www.ingramcontent.com/pod-product-compliance
Lightning Source LLC
Chambersburg PA
CBHW061334220326
41599CB00026B/5175